Greenkeeping Guide

A Beginner's Handbook for Healthy Turf

By Wallace Briggs

Authors Note

I have been working on golf courses for 36 years traveling from the rugged fairways of Alberta to the lush greens of British Columbia. Throughout my career, I have had the privilege of working at a remarkable golf course where two extraordinary players, Nick Taylor and Adam Hadwin, began their journeys from young talents to professional stars. Being a part of that environment taught me early on the profound impact that a well maintained course can have on developing raw talent. Over more than three decades in the industry, I've witnessed firsthand how meticulously cared for greens, innovative turf practices, and a commitment to environmental stewardship can elevate a golf course from a simple playing field to a sanctuary where the beauty and challenge of the game intertwine. My work has taken me through countless seasons of triumphs and challenges each experience adding to an ever growing toolkit of knowledge that I now share in this book. This guide is the culmination of my years of experience, blending tried and true traditional methods with modern innovations. It covers everything from the early mornings spent nurturing the perfect turf, and operational excellence. My journey has not only been about cultivating excellent playing surfaces but also about building relationships, learning continuously, and inspiring the next generation of golf course superintendents through writing this book. I wrote this book with a simple hope: that those who manage our cherished courses will find practical, actionable insights to help them cope with the challenges of modern turf management, while preserving the natural beauty that makes golf such a timeless sport. Whether you are just starting out or are a seasoned professional, I trust that my experiences from working the same grounds where legends began their careers to facing the dynamic challenges of modern management will resonate with you and help forge a path toward even greater success. Thank you for joining me on this journey. May these pages serve as both a practical guide and a reminder of the passion that drives our industry every day.

Printed Worldwide

Dedication

To the dreamers and custodians, to every young talent taking their first steps on the fairways, and to all the dedicated professionals who nurture that passion behind the scenes. This book is dedicated to those who transform every blade of grass into a foundation for excellence and who tirelessly work to create sustainable, stunning courses that inspire both play and possibility.

May these pages serve as a tribute to the passion and perseverance that drive our sport, and as a reminder that with every challenge comes the opportunity to build a brighter, greener future. Every swing, every step, and every decision you make paves the way for the next generation of greatness.

"Every blade of grass is a letter in the story of a course. With care, innovation, and a relentless commitment to excellence, we write a legacy that inspires every swing and nurtures the game we love."

Preface

In an industry where precision, innovation, and sustainability converge, the role of a golf course superintendent has evolved far beyond traditional turf maintenance. This book represents years of accumulated expertise, research, and practical insights designed to empower both new and seasoned professionals in the dynamic world of golf course management.

Many golf courses today face unprecedented challenges from managing complex environmental stressors and rising operational costs to meeting the growing demands for sustainable practices and digital transformation. At the same time, the advent of cutting edge automation, data-driven technologies, and advanced leadership models is reshaping the landscape of turf management. This book is your comprehensive guide to navigating these challenges and seizing the opportunities they present.

Throughout these pages, you will discover a detailed roadmap spanning the entire year from spring rejuvenation and summer maintenance to fall transition, winter planning, and beyond. Each chapter delves deeply into distinct yet interconnected areas:

Foundations and Planning: The early chapters lay the groundwork with robust budgeting, financial planning, and sustainability strategies, emphasizing the importance of proper course management as the backbone of long-term success.

Precision Turf Management: Explore advanced techniques such as GPS mapping, AI-driven diagnostics, and smart irrigation systems that are revolutionizing turf care. Here, technology meets tradition in a fusion that optimizes resource use, enhances turf quality, and reduces environmental impact.

Leadership and Business Strategies: Effective management goes hand in hand with strong leadership. Chapters on becoming an effective leader and developing sound business and budgeting strategies provide actionable insights into staff management, conflict resolution, and strategic financial planning that not only improve day to day operations but also lay the foundation for future growth.

Future Trends and Automation: As the industry looks ahead, robotics, autonomous systems, and advanced sensor technologies are poised to redefine operational practices. We examine these future trends in detail, weighing their transformative potential against real world economic considerations recognizing that while such innovations are currently within reach for high end courses, they serve as an aspirational benchmark for the industry as a whole.

Career Development and Networking: Success in golf course management is driven not only by technical skills but also by the strength of your professional network. In the final chapters, learn how to build your personal brand through certifications, industry conferences, and strategic relationship building with suppliers and partners.

This book is a culmination of meticulous research, detailed case studies, and practical checklists designed to serve as a reference guide for effective decision making. Whether you are looking to optimize your turf management processes, embrace sustainable practices, or spearhead a new era of operational efficiency through automation, the insights presented in these chapters will help you lead with confidence and shape the future of your organization.

Ultimately, this work is more than a technical manual it is a call to action for every golf course professional to embrace the dual imperatives of excellence and environmental stewardship. As you embark on this journey through each chapter, you will discover how small, incremental improvements can create monumental leaps in performance, profitability, and sustainability. The strategies you will learn here not only aim to improve the physical state of your turf but also to enhance the overall cultural, financial, and technological fabric of your organization.

Welcome to a new era in golf course management where every blade of grass, every drop of water, and every decision is informed by precision and driven by a commitment to excellence. May this book serve as your guide and inspiration as you navigate the present challenges and future opportunities of an evolving industry.

Chapter 1: The Role of a Golf Course Superintendent

Introduction

The role of a Golf Course Superintendent goes far beyond the day-to-day work of mowing greens and repairing bunkers. At first glance, many might assume that maintaining a golf course is as simple as keeping the grass trimmed and the bunkers raked. However, the true nature of the job is rich and multifaceted, blending deep expertise in agronomy with sophisticated business acumen, rigorous financial planning, and a strong set of leadership skills. It is a position that demands not only technical proficiency but also the ability to solve problems quickly, make well informed decisions under pressure, and manage a diverse team working toward a common goal.

A successful superintendent stands at the helm of course operations, shaping the overall experience that golfers enjoy. Whether it's enhancing the aesthetics and functionality of the playing surface or ensuring that every dollar is spent wisely, their day to day decisions have a direct impact on both the financial health of the course and the satisfaction of its patrons. To put it another way, while the visible effort may focus on the physical aspects of turf maintenance, the real engine behind the pristine greens and perfectly manicured fairways is a robust, well planned operation.

In this chapter, we delve into three fundamental aspects that define the role of a Golf Course Superintendent:

1. Budgeting and Financial Planning: Far more than just arranging for routine purchases or handling accounts, budgeting in this capacity involves a strategic approach where each dollar is an investment in the course's future. Superintendents must evaluate seasonal needs, anticipate unexpected challenges, and create a comprehensive financial roadmap that supports both the operational demands and long term improvements of the facility. This isn't just about balancing spreadsheets it's about aligning fiscal responsibility with the ultimate goal of stellar course performance.

2. Staff Training and Team Management: Behind the flawless appearance of any golf course is a dedicated team. The superintendent's effectiveness in training, mentoring, and managing this team directly influences the overall quality of the course. By establishing clear roles, fostering a collaborative atmosphere, and ensuring continuous professional development, they nurture an environment where every team member understands their part in a larger mission. This aspect of the role emphasizes both the human dimension of management and the practical need to have a reliable, well coordinated workforce that can respond to daily operational challenges.

3. Leadership Skills and Decision Making: Decision making in the field of golf course management isn't a task for the fainthearted. Whether it's responding to sudden weather changes, dealing with unexpected budget shortfalls, or navigating the intricacies of staff dynamics, a superintendent must exercise judgment that is both swift and prudent. This section of the chapter will illustrate how leadership isn't confined to issuing orders; it's about spearheading initiatives, setting a clear strategic vision, and continuously adapting to evolving circumstances always with an eye toward long term success.

Bridging the Technical and the Managerial

What makes the role of a Golf Course Superintendent particularly unique is the seamless integration of highly technical, hands on skills with broader managerial responsibilities. On any given day, a superintendent might switch from analyzing nutrient schedules to reviewing quarterly financial statements; from coaching a seasonal employee on proper equipment operation to re strategizing the maintenance schedule in response to an unexpected downpour. This duality requires a mindset that is as analytically rigorous as it is creative, as data driven as it is people oriented.

The interplay between these different facets of the role is critical. For instance, a sound financial plan can provide the resources necessary to invest in cutting edge irrigation systems, which in turn benefit turf health and reduce long term maintenance expenses. Similarly, a motivated and well trained team is more likely to execute complex tasks efficiently, thereby freeing up the superintendent's time to focus on strategic initiatives such as exploring new sustainable practices or integrating innovative technologies that enhance course performance.

The Broad Impact of a Superintendent's Work

Beyond the operational aspects, a superintendent's role has a broader impact on the golf course community and even on the local economy. A course that is well maintained not only attracts more golfers but also elevates the reputation of the club, leading to higher membership retention and an increase in event bookings. Moreover, by implementing environmentally sustainable practices, superintendents contribute to the larger goal of responsible land stewardship balancing the desires of avid golfers with the need to preserve natural resources for future generations.

In this light, the role transforms from merely a job description into a mission that ties together community engagement, environmental sustainability, and high standard operational excellence. Every decision made from the smallest adjustment in daily operations to the overarching financial strategy has a ripple effect that can influence the course's legacy for years to come.

Setting the Stage for What's to Come

As we embark on this exploration of the superintendent's domain, consider this chapter an invitation to immerse yourself in the strategic and operational intricacies that define the role. We will examine detailed methodologies for financial planning, delve into best practices for building and managing effective teams, and explore the decision making frameworks that empower leaders to tackle unforeseen challenges confidently.

In the pages that follow, you can expect to find in depth discussions, actionable strategies, and practical checklists designed to serve as both a guide and a reference. Whether you are an aspiring superintendent, a seasoned professional, or a stakeholder looking to understand the mechanics behind course management, this chapter lays the foundation for a comprehensive understanding of the role's multifarious demands.

Ultimately, the journey through this chapter and indeed, the entire book highlights the essence of what it means to be at the forefront of golf course management. It is a role that calls for intelligence, resilience, and a relentless commitment to excellence; one that transforms the seemingly mundane tasks of maintenance into strategic initiatives that ensure the course remains a timeless symbol of beauty, efficiency, and innovation.

Budgeting and Financial Planning: Building the Foundation

Understanding Your Financial Landscape

At its core, effective financial management for a golf course is about aligning expenditures with strategic objectives. For many superintendents, the annual budget isn't merely a ledger, but a dynamic plan that shapes every aspect of course maintenance and player experience. This begins with understanding the four major budget categories:

Turf Maintenance Costs: This includes fertilizers, pesticides, seed mixtures, irrigation expenses, aeration, and over seeding. Since these costs can vary significantly by season, careful planning protects the integrity and playability of the turf.

Labor & Staffing: Staffing costs encompass not only wages and benefits for full time employees but also expenses for seasonal workers, specialized consultants, and training initiatives. A well planned staffing budget ensures that the right talent is in place during peak seasons and that proper training enhances productivity.

Equipment & Infrastructure: Investments in mowers, aerators, irrigation systems, and other critical equipment are essential. Routine service contracts, repairs, and occasional technological upgrades factor into this category.

Operational Overheads & Contingency: Utility bills, fuel costs, routine upkeep for paths and bunkers, and unexpected repairs belong here. It is wise to set aside a dedicated emergency fund (typically 10 to 15% of the budget) to manage unforeseen occurrences such as extreme weather, disease outbreaks, or sudden equipment failures.

Diagram Idea: Imagine a pie chart depicting the percentage distribution among these categories, with annotations showing typical seasonal variations. Such a visual can easily be incorporated into the chapter to give readers a quick snapshot of where funds must be allocated throughout the year.

Developing an Annual Budget Cycle

A golf course's operating budget is inherently seasonal. Superintendents face differing financial demands through the year:

Spring: Marked by a burst of activity, spring sees significant spending on aeration, over seeding, and applying the first rounds of fertilizers and pesticides.

Staff onboarding and training are crucial at this stage, meaning labor costs rise.

Summer: As the course reaches prime playing conditions, summer demands increase in labor to maintain rapidly growing turf, along with expanded irrigation costs. The focus is on sustaining turf health during heat stress.

Fall: With the onset of cooler weather, attention shifts toward preparing the turf for dormancy. Budget items here include over seeding weak areas, deep aeration, and treatments for disease prevention.

Winter: Although the playing surface is less active, winter budgets are still busy with equipment repairs and maintenance. Reduced staff hours and planning for the next cycle dominate this period.

Real World Example: One superintendent noted that "By tracking seasonal spending patterns over several years, we were able to allocate more funds in the spring and summer, thereby reducing our emergency spending in the fall." Such insights illustrate how historical data informs more effective budgeting.

Balancing Cost and Quality

An effective superintendent always seeks a balance between cost control and quality maintenance. Several strategies can help achieve this balance:

Prioritize Investments: For instance, investing in premium soil amendments can reduce the need for frequent pesticide or fungicide treatments later. Long term savings and improved turf resilience are often worth the initial higher expense.

Negotiation and Bulk Purchasing: Forming strong relationships with suppliers can yield discounts and favorable payment terms. Many superintendents negotiate multiyear contracts to lock in current prices and budget stability.

Adopt Sustainable Practices: Integrated Pest Management (IPM) and smart irrigation systems not only support environmental stewardship but also reduce recurring costs. Adopting such practices can be presented as case studies within the chapter.

Leveraging Data and Technology

In today's digital age, technology is playing an increasingly pivotal role in budgeting and financial planning. Some methods include:

Automated Financial Software: Tools that track monthly expenses, forecast seasonal demands, and alert superintendents to unusual spending can vastly improve budgeting accuracy.

GPS Based Turf Monitoring: By employing GPS and mapping software, superintendents can identify areas requiring concentrated attention, thereby avoiding the wasteful application of resources.

Data Analytics: Regular review of historical spending and performance metrics allows for more informed financial decision making in the future.

Checklist Budgeting Essentials:

Draft a detailed annual budget covering all key expense categories.

Allocate funds based on seasonal needs and historical spending patterns.

Maintain an emergency fund (10 15% of the total budget) for unexpected incidents.

Invest in technology to track and analyze spending.

Regularly review and adjust the budget based on performance data.

Staff Training and Team Management: Building Your Workforce

The Importance of a Skilled Team

A golf course's quality is directly tied to the expertise and dedication of its staff. The superintendent must cultivate a team that is not only well trained but also motivated to uphold the highest standards. This begins with a clear understanding of each team member's role, responsibilities, and contribution to course upkeep.

Structuring the Team

Effective team management starts with clear role assignment. A successful team might include roles such as:

Greenkeepers: Responsible for the upkeep of greens, including mowing frequencies, aeration, and pest management.

Fairway and Rough Crew: Focused on maintaining the quality of fairways and surrounding areas, ensuring consistent playability.

Irrigation Specialists: Technicians who manage the operation, repair, and optimization of irrigation systems.

Seasonal Workers: Supplementary staff brought in during high activity periods, who require rapid, comprehensive training.

Diagram Idea: A flowchart showing the organizational hierarchy can help visualize reporting lines and responsibilities from the superintendent down through team leads and individual workers. This diagram can be a valuable reference for supervisors seeking clarity in delegation.

Training and Onboarding: Setting the Standard

Given that many courses rely on a mix of full time and seasonal staff, the onboarding process is critical. Here's how to ensure that every new employee is aligned with the course's operational standards:

Orientation Sessions: Begin with a comprehensive overview of the golf course's history, operational priorities, and the standards expected of every staff member.

Hands on Training: Practical demonstrations of equipment use, turf management techniques, and safety protocols are essential. For example, having seasonal workers shadow experienced staff members during busy periods can exponentially increase their on ground readiness.

Regular Workshops: Schedule workshops focusing on the latest trends in turf management, new technology deployments, and recommended practices. This continuous training helps all levels of staff stay updated with the evolving challenges of course maintenance.

Safety Protocols: Given the risks associated with operating heavy machinery and handling chemicals, regular safety briefings and certification programs are non negotiable.

Case Example: A well known course in the Pacific Northwest instituted a quarterly training program where staff members rotated through different operational areas. This cross training not only improved overall department flexibility but also fostered a sense of shared responsibility, leading to better job satisfaction and higher efficiency.

Cultivating a Positive Work Environment

Beyond technical training, managing team dynamics is equally important. A positive work environment translates to better performance and lower turnover. Consider the following strategies:

Recognition and Incentives: Regularly acknowledge exceptional work with awards, bonus incentives, or public recognition during team meetings.

Open Communication: Create channels for feedback where staff can share ideas and concerns. This openness helps in addressing issues promptly and builds mutual respect.

Team Building Activities: Organize off site or on site events that encourage bonding and improve morale. Whether it's a seasonal picnic or a team challenge, mutual camaraderie is central to achieving peak performance.

Professional Development: Provide opportunities for career advancement, such as sending promising staff to industry conferences or certification courses. Demonstrating investment in an employee's growth often leads to a more dedicated workforce.

Checklist Team Management Essentials:

Establish clear job descriptions for every team member.

Develop and document an onboarding process and training curriculum.

Schedule regular safety and skills workshops.

Implement a feedback mechanism and recognition program.

Encourage cross training to foster versatility within the team.

Leadership Skills and Decision Making: Guiding with Confidence

Defining Leadership in Course Superintendency

Being a successful Golf Course Superintendent means being a leader first a strategic thinker who not only directs daily activities but also plans for long term successes. Leadership in this context involves several intertwined qualities:

Visionary Planning: The ability to foresee future challenges and identify opportunities for improvement. This forward thinking mindset is critical for sustainable course management.

Effective Communication: Whether coordinating with staff, reporting to owners, or interacting with the golfing public, communication skills are paramount. Clear, concise, and empathetic dialogue helps maintain transparency and foster trust.

Problem Solving: Unexpected challenges are inevitable. Whether facing sudden weather anomalies, equipment breakdowns, or staffing shortages, a superintendent must diagnose problems quickly and implement effective remedies.

Adaptability: The industry evolves with new technologies and changing environmental standards. Staying current on research, techniques, and industry trend is a hallmark of an effective leader.

Developing Decision Making Skills

Decision making in this role isn't just about reacting swiftly; it requires a structured approach. Consider the following framework:

1. **Gather Data:** Whether through regular financial reports, turf health readings, or staff feedback, make informed decisions by collecting accurate and up to date information.

2. **Analyze Options:** Use historical data and present conditions to evaluate multiple courses of action. For example, if a sudden pest outbreak occurs, weigh the costs and benefits of immediate chemical treatments versus preventative measures implemented earlier in the season.

3. **Consult with Experts:** Leverage the collective expertise of your team. A quick discussion with a senior greenkeeper or an irrigation specialist might reveal insights that you may have overlooked.

4. **Implement with Flexibility:** After choosing a course of action, remain flexible. Monitor results closely and be ready to recalibrate if conditions deteriorate or unexpected outcomes arise.

Leadership in Action: Real World Strategies

Let's explore a few scenarios that illustrate effective leadership:

Weather Related Challenges: During a period of sudden heavy rains, a superintendent noticed increasing waterlogging on greens. By quickly reallocating resources to improve drainage systems and adjusting the mowing schedule, he prevented long term turf damage. His decision was based on data collected via moisture sensors and proactive communication with staff.

Handling Staff Shortages: In one case, a key team member was unexpectedly absent during a critical period. Rather than simply redistributing tasks, the superintendent held a brief team meeting, re assigned responsibilities based on current skills, and provided immediate onsite training. This not only ensured continuity of work but also boosted overall team morale.

Balancing Budget Constraints: When faced with cost cutting measures, a superintendent opted to invest in a higher quality, more efficient irrigation system. Although the upfront costs were higher, the long term savings in water bills and reduced maintenance expenditures quickly offset the investment. This decision was made after negotiating with suppliers, reviewing energy consumption data, and consulting environmental experts.

Diagram Idea: A decision making flowchart can be a powerful tool. Imagine a diagram that starts with "Identify Issue" and branches into possible responses, gathering information at each node until arriving at an "Implement Solution" box. Such a visual not only clarifies the process but also serves as a reference tool for new superintendents.

Fostering a Culture of Continuous Improvement

Effective leadership is also about nurturing a culture that embraces continuous improvement. Consider these leadership practices:

Set Clear Objectives: Outline quarterly performance goals for both the turf and the team, using measurable benchmarks.

Review and Reflect: At the end of each season, hold a debrief session with your core team to discuss what worked, what didn't, and what improvements can be made.

Encourage Innovation: Invite team members to suggest new methods or technologies that could enhance course operations.

Lead by Example: Demonstrate dedication and resilience. When your team sees you working through challenges and staying committed to excellence, they too will adopt these qualities.

Checklist Leadership Essentials:

Establish a clear, strategic vision for the season and beyond.

Develop a reliable system for data collection and analysis.

Create an open forum for feedback and innovation.

Monitor outcomes and adjust strategies based on performance data.

Invest in your own professional development through industry conferences, readings, and mentorship opportunities.

Synthesis and Action

By integrating effective budgeting, robust team management, and visionary leadership, a Golf Course Superintendent creates an environment where operational excellence and financial prudence coexist with the highest quality of turf care. This comprehensive approach not only elevates the overall playing experience but also ensures the course stands the test of time amid evolving challenges.

Putting It All Together: A Day in the Life Scenario

Imagine a typical early spring morning on a well run course. The superintendent reviews the latest financial reports on a tablet, noting that the emergency fund remains robust after recent heavy rains a result of last year's careful budget allocations. Following this, a quick briefing with the team reinforces the day's priorities: scheduled aeration in high traffic areas and targeted over seeding in sections that suffered from winter damage.

Later that day, issues with irrigation in a newly seeded green prompt a singularly focused problem solving session. Drawing on real time sensor data and insights from the irrigation specialist, the superintendent quickly adjusts system parameters and sets a follow up meeting to review long term performance improvements. Through these decisions, the superintendent not only preserves turf health but also demonstrates how data, collaboration, and proactive planning converge into sound decision making.

In the afternoon, during a routine walk across the course, the superintendent notices an emerging pattern of drought stress in a remote area of the fairway. Using this opportunity, she calls for a quick team huddle, delegate's immediate corrective measures, and later updates the digital dashboard with lessons learned for future reference. These actions highlight the interplay of financial oversight, teamwork, and leadership within the framework of day to day operations.

Conclusion

The multifaceted role of a Golf Course Superintendent requires deep expertise in three critical domains: financial stewardship, team development, and leadership. By diligently managing budgets, fostering a skilled workforce, and making informed, strategic decisions, superintendents ensure that every blade of grass contributes to both the beauty of the course and its operational efficiency.

This chapter has provided an in depth look at:

Budgeting and Financial Planning: The development of annual budgets, seasonal financial planning, cost quality tradeoffs, and the leverage of modern technology.

Staff Training and Team Management: Comprehensive onboarding processes, effective team structuring, staff motivation, and continuous training.

Leadership Skills and Decision Making: The formulation of a visionary strategy, the utilization of data driven decision frameworks, handling unexpected challenges, and the cultivation of a culture of constant improvement.

By integrating these areas, superintendents not only safeguard the financial health of the golf course but also elevate the overall experience for staff and golfers alike. Whether facing unpredictable weather, staff challenges, or budgetary constraints, a proactive and thought out approach ensures the course continues to set industry benchmarks.

End of Chapter Checklist: Superintendent Essentials

Budgeting and Financial Planning:

Draft an annual budget that reflects seasonal demands.

Monitor expenditure trends and allocate funds for emergencies.

Leverage technology and data analytics for smoother financial management.

Negotiate supplier contracts to maximize cost efficiencies.

Staff Training and Team Management:

Develop detailed job descriptions and thorough onboarding processes.

Schedule regular safety and skills workshops for all team members.

Establish clear lines of responsibility and encourage cross training.

Implement feedback systems and recognition programs to maintain high morale.

Leadership Skills and Decision Making:

Set clear strategic objectives and review them on a seasonal basis.

Create a structured framework for rapid problem solving.

Facilitate open communication channels within the team.

Regularly attend professional development sessions and industry conferences.

Final Thoughts

In this chapter, we've laid the groundwork by exploring how financial acumen, effective team management, and strong leadership form the backbone of a successful golf course operation. The insights, checklists, and diagram ideas provided here aim to guide you as you take on the role of a Golf

Course Superintendent ensuring every decision contributes to a thriving, well managed course. As we progress through this book, further chapters will build upon these foundational concepts, deepening your practical toolkit and elevating your approach to course management.

Chapter 2: Turf grass Science & Course Design

Golf courses are living, evolving landscapes that require a nuanced understanding of turf grass science and thoughtful design to balance aesthetics, playability, and environmental sustainability. The foundation of a great course starts below the surface with the soil; continues through the lifelines of water via the irrigation system; and is refined by the rhythmic precision of mowing practices. This chapter explores these three pillars, revealing the technical and practical insights needed to nurture lush, healthy turf that meets both performance standards and aesthetic demands. Superintendents who master these disciplines can greatly enhance the quality of the course, reduce long term maintenance costs, and ensure the sustainability of their grounds.

In what follows, we will delve deeply into:

1. Soil Testing and Nutrient Management: Understanding your soil is the first step to creating and maintaining an ideal turf environment.

2. Irrigation System Checks and Improvements: Water is the lifeblood of turf health. We'll explore how to optimize water use while preserving quality.

3. Mowing Strategies for Optimal Turf Growth: Precision mowing not only shapes the course but also plays a vital role in the biological health of the turf.

Each section includes detailed explanations, real world applications, diagram concepts, and checklists for quick reference and implementation.

Soil Testing and Nutrient Management

The Foundation Begins Beneath the Surface

Every great golf course starts with healthy soil. Soil testing is the critical first step in understanding the nutrient profile; pH levels, organic matter content, and microbial activity that directly influence turf performance. Advanced soil testing methods, including core sampling and laboratory analysis, provide insights that bridge the gap between the biological needs of turf grass and the management practices that support it. By regularly testing your soil, you can make informed decisions about fertilizer application, the use of soil amendments, and long term nutrient management strategies.

Methods and Best Practices for Soil Testing

Sampling Technique:

Core Sampling: Use a soil core sampler to extract samples from representative areas of the course. It's crucial to take samples from various zones (greens, fairways, tees, and roughs), as each area may exhibit different soil characteristics.

Composite Sampling: Combine several samples from similar areas to create a composite sample. This helps to smooth out anomalies and achieve an accurate overall picture of the soil's condition.

Analyzing pH and Nutrient Levels:

PH Balance: The pH of the soil affects nutrient availability. Most turf grasses prefer a pH between 6.0 and 7.0. Adjustments can be made with lime (to raise pH) or sulfur (to lower pH).

Macro and Micronutrients: Test for primary nutrients like nitrogen (N), phosphorus (P), and potassium (K), as well as secondary nutrients and micronutrients. Identifying deficiencies early on helps prevent turf stress during peak growing periods.

Frequency and Timing:

Seasonal Testing: Conduct soil tests at the beginning of each major season, especially during the transition from dormancy in spring and before the onset of summer stress. Consistent testing (annually or biannually) provides a baseline over time, enabling you to track improvements or emerging problems.

Technology Integration: Utilize digital pH meters and portable testing kits for on the spot readings during routine inspections. Advanced laboratories may also offer rapid turnaround tests for microbial activity that informs organic management practices.

Developing a Nutrient Management Plan

Once the soil test results are in, the next step is developing a comprehensive nutrient management plan. This plan should tie closely to the nutrient needs of the turf and the ecological dynamics of the soil.

Setting Fertilizer Goals:

Based on the results, determine the type and rate of fertilizers needed for each section of the course. Consider using slow release fertilizers to provide a steady supply of nutrients and reduce the chance of nutrient leaching.

Timing and Application: Plan application schedules that align with turf growth cycles. For example, pre planting fertilizers in early spring can support initial growth rapidly, while maintenance fertilizations through summer and fall help sustain healthy turf during peak play and recovery periods.

Organic and Integrated Practices:

Organic Matter: Incorporate compost or other organic amendments to boost soil structure and water retention, particularly in areas with sandy or compacted soils.

Integrated Pest and Nutrient Management (IPNM): Combine nutrient applications with pest management strategies. A balanced microbial ecosystem in the soil can help naturally ward off diseases, reducing reliance on chemical treatments.

Monitoring and Adjustments:

Regularly review the results of your soil tests against the outcomes in turf performance. Adjust fertilizer applications as needed, and maintain records to create a long term nutrient profile for every section of the course.

Diagram Idea: Imagine a chart or flow diagram that illustrates the steps of soil testing from sample collection and laboratory analysis to the interpretation of results and formulation of a nutrient management plan. Such a visual tool can be included in this section to provide an at a glance overview for busy superintendents.

Checklist Soil Testing and Nutrient Management Essentials

Develop a sampling plan that includes all key areas (greens, fairways, tees, roughs).

Conduct composite sampling for consistent results.

Test for pH, macro nutrients (N P K), and key micronutrients.

Implement a routine (seasonal/annual) soil testing schedule.

Create a nutrient management plan that aligns with seasonal turf needs.

Track changes in soil health and adjust fertilizer use accordingly.

By embracing these practices, superintendents can create a proactive approach to soil health a vital underpinning for every other turf management decision.

Irrigation System Checks and Improvements

The Crucial Role of Water in Turf Health

Water is the lifeblood of any golf course, and a well designed irrigation system is essential for maintaining high quality turf under various weather conditions. Beyond simply delivering water, a modern irrigation system must be efficient, precise, and responsive to the nuanced needs of different areas of the course. From high tech sensors to traditional schedule based systems, understanding and maintaining your irrigation infrastructure directly affects both water conservation and turf health.

Components of a Robust Irrigation System

Distribution Networks and Controls:

Sprinkler Heads and Nozzles: Different areas of a course may require different types of sprinklers from high volume rotaries for fairways to precision nozzles for greens. Regular checks for clogs, wear, and misalignment are crucial.

Control Systems: Modern irrigation systems use automated controllers that can be programmed to adjust watering schedules based on real time weather data. Investing in advanced controllers can yield significant water savings and ensure consistent turf moisture.

Pressure Regulation and Zones:

Pressure Checks: Consistent water pressure is necessary to ensure uniform distribution across all zones. Pressure regulators and valves should be inspected and calibrated periodically.

Zoning: Dividing the course into multiple irrigation zones allows for tailored watering, addressing the specific needs of greens, tees, fairways, and roughs. Each zone can be managed based on exposure, soil type, and sun intensity, resulting in optimal water use.

Regular System Checks and Preventive Maintenance

To ensure the system operates at peak efficiency, superintendents should schedule routine inspections:

Pre Season and Post Season Inspections:

Pre Season: Before the start of the major playing season, inspect every irrigation zone for signs of leaks, clogged nozzles, or faulty controllers. Establish a baseline performance for the entire system.

Post Season: After the heavy use periods of summer and fall, evaluate the system's performance to identify any wear or degradation that occurred during peak usage. Minor repairs and adjustments made during off peak times can prevent costly problems during the busy season.

On Demand Checks:

Weather Triggered Inspections: After severe weather events such as heavy rains or unexpected winds inspect the system promptly to ensure no damage has occurred. Storms can dislodge sprinkler heads, damage piping, or blow debris into critical components.

Scheduled Routine Checks: Incorporate short, routine inspections (weekly or biweekly, depending on the season) to check pressure gauges, controller logs, and the physical condition of equipment.

Technology for Efficiency:

Smart Irrigation Controllers: Utilize systems that integrate real time weather data and soil moisture sensors. These tools automatically adjust watering times to conserve water while ensuring adequate turf hydration.

Mapping and Data Analytics: Some modern irrigation systems allow for mapping of water distribution, highlighting over watered or under watered areas. Analyzing this data enables superintendents to fine tune system performance over time.

Upgrading and Improving the System

Irrigation system improvements can yield long term benefits in water savings and turf quality. Consider the following steps for system upgrade:

Retrofitting Existing Systems:

Upgrade outdated controllers with programmable, sensor driven models.

Replace worn out sprinkler heads with energy efficient and clog resistant designs.

Redefine irrigation zones based on recent data and turf observations, ensuring each zone receives the optimal amount of water.

Investing in New Technology:

Evaluate the cost benefit of installing drip irrigation components in areas with particularly poor water retention or unique growth challenges.

Leverage developments in remote monitoring and irrigation management software to monitor and control the system from a central, digital platform.

Diagram Idea: A flow diagram or schematic showing the irrigation system layout can help visualize the network of pipes, valves, and controllers. Such a diagram can include recommended check points, highlighting where sensors should be placed and where pressure tests are conducted.

Checklist Irrigation System Maintenance Essentials

Schedule pre season, post season, and routine inspections for all irrigation zones.

Regularly clean and test sprinkler heads, nozzles, and controllers.

Verify water pressure in each zone and adjust regulators as needed.

Utilize smart controllers and moisture sensors to optimize watering schedules.

Document system performance and plan upgrades based on data analytics.

Quickly address any issues following severe weather events.

By maintaining a vigilant approach to irrigation system checks and improvements, superintendents can maximize water efficiency, reduce maintenance costs, and foster lush, resilient turf.

Mowing Strategies for Optimal Turf Growth

The Art and Science of Mowing

While soil testing and irrigation supply the foundational inputs for turf health, the practice of mowing shapes that health into a form suitable for play. Mowing is more than a landscaping chore it's an integral technique that influences turf density, growth patterns, and overall resilience. Employing the right mowing strategy is essential to prevent diseases, reduce wear, and ensure a smooth, aesthetically pleasing playing surface.

Techniques for Effective Mowing

Mowing Frequency and Timing:

Regular Intervals: Mowing at consistent intervals promotes uniform turf growth and reduces the risk of disease. The optimal frequency depends on the grass type, growth rate, and season. For example, during peak growing periods in spring and early summer, more frequent mowing is required to maintain the optimal height.

Time of Day: Early morning or late afternoon mowing is recommended to avoid midday heat stress, reduce evaporation losses, and minimize turf damage. These practices help ensure that the grass is not cut too short under conditions that may exacerbate stress.

Mowing Height and Patterns:

Adjusting Mower Height: Different parts of the golf course require different mowing heights. Greens typically require a lower cut (1/8 to 1/4 inch) to achieve speed and smoothness, while fairways and roughs can be maintained at a higher height to promote resilience and reduce stress.

Pattern Rotation: Changing mowing patterns periodically (e.g., alternating between linear, diagonal, or circular patterns) prevents the turf from developing a "memory" of the cut and helps reduce wear in specific areas. This also minimizes soil compaction by dispersing foot traffic more evenly.

Equipment Maintenance and Mower Upkeep:

Blade Sharpness: Dull blades can stress the turf by tearing rather than cutting the grass cleanly, thereby increasing the risk of disease and uneven growth. Regularly sharpening or replacing blades is essential.

Mower Calibration: Ensure that mowers are correctly adjusted for cutting height and that the equipment is well maintained to run smoothly. Routine maintenance, including cleaning the deck and checking for wear, supports both the health of the turf and worker safety.

Technology Integration: Consider investing in precision mowing equipment equipped with height sensors or automated settings that adapt to changing turf conditions. These features can help maintain optimum mowing conditions consistently.

Developing a Mowing Plan

Creating a tailored mowing plan involves understanding your turf's unique requirements and adapting strategies based on seasonal growth patterns:

Seasonal Adjustments:

Spring and Early Summer: Focus on frequent, lighter cuts to encourage lush, vigorous growth. Monitor weather conditions, as wet conditions may require slower mowing speeds to avoid slipping or turf damage.

Mid summer: During periods of heat stress, adjust mowing frequencies and heights to minimize stress. Raise mowing heights slightly to protect the grass from intense sun exposure.

Fall and Dormancy: As the growing season winds down, reduce mowing frequency and increase cutting heights. This allows older, denser growth to cushion the impacts of cooler weather and conserve energy.

Leveraging Observational Data:

Visual Assessments: Regularly inspect the turf for signs of stress, uneven growth, or disease. Adjust mowing practices accordingly if certain areas are prone to thatch buildup or compaction, consider alternative mowing patterns or additional aeration.

Documentation: Maintain records of mowing patterns, frequencies, and any observed effects on turf health. This historical data can be invaluable for fine tuning future mowing plans and integrating lessons learned into routine practices.

Diagram Idea: A diagram illustrating different mowing patterns (e.g., straight lines, crisscross, diagonal) can be an excellent visual aid. Such a tool not only demonstrates the techniques but also shows the benefits of rotating patterns to minimize turf stress.

Checklist Mowing Strategy Essentials

Establish a regular mowing schedule adapted to seasonal growth patterns.

Maintain and sharpen mower blades consistently to ensure clean cuts.

Adjust cutting heights for different areas: lower on greens, higher on fairways and roughs.

Rotate mowing patterns regularly to reduce compaction and promote uniform growth.

Document mowing practices and observe turf responses to inform future adjustments.

Conduct routine equipment checks and calibrations to ensure consistent performance.

Through disciplined, informed mowing practices, superintendents can create a visually appealing course that also optimizes turf health and longevity. The right balance between frequency, height, and technique translates directly into improved playability and reduced long term maintenance needs.

Conclusion

Turf grass science and course design are the heart and soul of a well managed golf course. This chapter has provided a deep exploration of three critical areas:

Soil Testing and Nutrient Management: Laying the foundation by understanding your soil's composition, implementing consistent testing protocols, and developing nutrient strategies that respond to the course's unique needs.

Irrigation System Checks and Improvements: Highlighting the importance of an efficient, technology driven irrigation system that delivers water precisely where and when it's needed, while incorporating regular maintenance and upgrades.

Mowing Strategies for Optimal Turf Growth: Revealing how meticulous mowing practices tailored to seasonal conditions and executed with high quality equipment contribute significantly to turf health and overall course quality.

By integrating these practices, golf course superintendents can elevate the quality of their turf, optimize maintenance practices, and create a sustainable environment that meets both aesthetic and performance goals. The detailed checklists, diagram ideas, and real world applications provided in this chapter serve as practical tools that readers can immediately apply in their daily operations.

As we continue through this book, the focus will shift toward additional technical and managerial aspects of turf maintenance and course design, building upon the robust foundation established here. With a solid understanding of soil, water, and mowing basics, you are well equipped to tackle the challenges and opportunities that lie ahead in creating and maintaining a world class golf course.

This comprehensive chapter not only imparts technical knowledge but also sets the stage for the continuous, strategic improvements necessary in modern course management. It serves as an invaluable resource for both new enthusiasts and seasoned professionals aiming to elevate their turf management practices to the next level.

Chapter 3: Equipment & Infrastructure Maintenance

A well run golf course is defined not only by its pristine turf and innovative turf management practices but also by the reliability of its physical infrastructure and equipment. From tractors and mowers that shape the greens to the smooth cart paths that contribute to the overall visitor experience, every element demands regular care and strategic planning. In this chapter, we bring into focus the importance of routine equipment servicing, enhancing frequently traveled areas such as cart paths and clubhouse landscaping, and setting up robust preventative maintenance schedules that keep operations running without unexpected breakdowns.

The interplay between proper equipment care and timely maintenance of infrastructure underpins the operational smoothness of a course. With high value investments in machinery and physical assets, ensuring their longevity is paramount not only to maintain appearance and performance but also to control costs over time.

Routine Equipment Servicing

The Importance of Regular Equipment Servicing

Modern golf course equipment ranging from mowers and aerators to specialized turf management implements is a significant investment critical for daily operations. Routine equipment servicing ensures that all machinery functions safely, efficiently, and reliably. Regular checkups help to detect early signs of wear and tear, prevent costly breakdowns, and extend the lifespan of your assets. This preventive care also enhances safety, reducing the chances of accidents on course and in maintenance areas.

Key Components of Equipment Servicing

A. Scheduled Inspections and Servicing Intervals

Daily Checks:

Visual Inspections: Every shift should begin with a visual assessment of equipment. Check for obvious signs of damage or wear, such as frayed cables, leaning attachments, and any leakage of fluids.

Functionality Tests: Basic operational tests ensure that engines start promptly, hydraulic systems work without chugging, and safety features (like emergency shutoffs) are active.

Weekly Maintenance:

Lubrication and Cleaning: Clean the equipment of dust, grass clippings, and debris that can accumulate. Lightly lubricate moving parts, check air filters, and clear out any blockages.

Fluid Levels: Inspect oil, coolant, and fuel levels. Recharge or replace fluids as necessary to maintain peak efficiency.

Monthly and Seasonal Servicing:

Detailed Inspections: On a monthly or seasonal basis, conduct detailed inspections that include engine diagnostics, belt tension checks, and complete overhauls of complex systems.

Calibration and Adjustments: Ensure that equipment such as mowers is calibrated correctly for cutting heights and blade alignment. Seasonal checks may also include tire pressure reviews on automated or self propelled mowers and assessments of hydraulic system performance.

B. Record Keeping and Performance Logs

Maintenance Logs: Keeping a detailed log of servicing schedules, parts replaced, and any adjustments made allow for trend analysis and anticipate future repairs.

Manufacturer Guidelines: Always consult the manufacturer's recommended service intervals and parts specifications to ensure compliance and the longevity of equipment warranties.

C. Safety and Training Protocols

Operator Training: Ensure that all equipment operators are trained not only in using the machinery but also in conducting basic maintenance checks themselves.

Personal Protective Equipment (PPE): Supervisors must remind operators to use key safety gear during servicing sessions, such as gloves, safety glasses, and proper footwear.

Real World Example

A superintendent in the Midwest shared that by implementing a daily pre operation checklist combined with a monthly comprehensive service schedule for all mowing equipment, they reduced unscheduled breakdowns by 40% over the course of a season. Investing in robust record keeping software further allowed the team to identify recurring issues and negotiate better service contracts with local providers.

Diagram Idea

Imagine a flowchart that outlines the servicing schedule:

Daily: Pre use check, Visual Inspection, Basic Function Test

Weekly: Cleaning, Lubrication, Fluid Check

Monthly: Detailed Diagnostics, Calibration, Log Update Such a diagram can serve as a quick reference tool for supervisory teams during the equipment servicing routine.

Checklist Routine Equipment Servicing Essentials

Begin each shift with a pre operation visual and functionality check.

Perform weekly cleaning and lubrication of moving parts.

Check fluid levels oil, coolant, and fuel regularly.

Conduct detailed monthly inspections including calibration and belt tension checks.

Maintain comprehensive maintenance logs with dates, actions taken, and any parts replaced.

Ensure all operators are equipped with and trained in the proper use of PPE during maintenance.

By adhering closely to these routines, superintendents not only safeguard the safety of their teams but also ensure that equipment keeps pace with the rigorous demands of course maintenance.

Cart Path and Clubhouse Landscaping

Enhancing the Visitor Experience Through Infrastructure

While the focus of golf course maintenance is often on turf and greens, the outward facing infrastructure especially cart paths and clubhouse landscaping plays an equally crucial role in creating an excellent visitor experience. Well maintained cart paths ensure smooth travel across the course, while immaculate clubhouse landscaping enhances the course's overall aesthetic appeal and reinforces the brand perception of quality.

Maintaining and Upgrading Cart Paths

A. Routine Inspections and Repairs

Surface Assessments: Regularly inspect cart paths for cracks, potholes, and areas of erosion. Frequent foot and cart traffic, along with weather induced wear and tear, can degrade the path surface.

Drainage and Water Runoff: Ensure that cart paths have adequate drainage systems to prevent water pooling, which can accelerate decay.

Material Rejuvenation: Depending on the path's construction be it concrete, asphalt, or composite materials routine resealing and surface treatments can extend the lifespan of the paths.

B. Upgrading Technologies and Materials

New Materials: Advances in materials offer more durable, weather resistant surfaces. Innovations such as recycled rubber overlays or polymer modified sealants can provide increased traction and longevity.

Regular Maintenance Contracts: Partnering with specialized contractors for periodic resurfacing or deep cleaning can be a cost effective solution in the long term.

Clubhouse Landscaping: Creating a Welcoming Environment

A. Comprehensive Landscape Design

Exterior Aesthetics: The clubhouse exterior should reflect the course's overall style and quality. This includes well kept lawns, flower beds, and strategically placed trees and shrubs.

Seasonal Displays: Consider planting seasonal colors or themed gardens that change throughout the year, which can elevate the visual appeal and create a welcoming atmosphere for players and guests.

B. Maintenance Practices for Landscaping

Routine Care: Schedule regular lawn mowing, pruning, and mulching. High quality landscape maintenance doesn't happen by accident it is driven by regular, consistent care.

Irrigation and Fertilization: Just as with turf, the landscaping areas around the clubhouse benefit from carefully managed irrigation and nutrient programs.

Pathways and Signage: Maintain the aesthetic quality of walkways leading to and around the clubhouse, ensuring that they are free of debris, cracks, or unsightly repairs. Strategic placement of signage and decorative elements can further enhance the overall appeal.

Real World Example

A golf course in the Pacific Northwest integrated a quarterly review of their cart path conditions with a targeted maintenance plan. They discovered that a combination of heat and water runoff was causing premature surface wear. By installing additional drainage and upgrading the sealing material every two years, they extended the lifespan of the cart paths by an estimated 50%. Similarly, regular updates to the landscaping around the clubhouse with seasonal flower displays and consistent pruning resulted in a 20% increase in positive member feedback regarding the course's overall ambiance.

Diagram Idea

Consider a before and after visual narrative: a diagram that shows a cross section of a cart path with proper drainage, then a map of the clubhouse with key landscaping elements (flower beds, erosion control areas, designated seating, and aesthetic lighting). This not only demonstrates the technical aspects of maintenance but also the transformational impact on the visitor experience.

Checklist Cart Path and Clubhouse Landscaping Essentials

Conduct monthly inspections of cart paths for cracks, potholes, and drainage issues.

Schedule seasonal resurfacing and sealing based on the material used.

Ensure proper drainage is maintained along all pathways.

Plan for seasonal landscaping updates around the clubhouse.

Maintain regular pruning, mowing, and fertilization schedules for exterior grounds.

Regularly clear debris and inspect walkways to maintain a welcoming, safe environment.

With strategic attention to cart path repair and clubhouse landscaping, a golf course enhances its first impression and overall usability key factors in attracting and retaining members and visitors.

Preventative Maintenance Schedules

The Lifeblood of Long Term Operational Success

Preventative maintenance is the proactive approach to repair and upkeep that minimizes unexpected equipment failures, extends the life of assets, and avoids costly emergency repairs. By establishing systematic schedules and adhering to them rigorously, superintendents can optimize operational turnout and avoid the pitfalls and downtime associated with reactive maintenance protocols.

Key Components of Preventative Maintenance

A. Scheduled Infrastructure Inspections

Comprehensive Roadmaps: Develop an annual calendar that outlines maintenance tasks ranging from equipment servicing to infrastructure repairs. Each calendar should detail critical checkpoints for elevated risks such as seasonal component wear.

Systematic Reviews: Incorporate periodic reviews of each piece of equipment and infrastructure component. For example, biannual deep inspections may be scheduled for high use machinery, while monthly spot checks might suffice for less critical systems.

B. Technology Driven Monitoring

Remote Monitoring Systems: Advances in IoT (Internet of Things) technology now allow superintendents to monitor equipment performance and infrastructure conditions remotely. Sensors can alert managers to issues such as overheating, abnormal vibration, or moisture intrusion.

Data Analytics: Software can consolidate maintenance logs and sensor outputs, enabling predictive maintenance algorithms to forecast when a component is likely to fail. This allows for repairs to be planned well before disruptions occur.

C. Workforce Coordination and Training

Cross Functional Teams: Preventative maintenance is most effective when it's a shared responsibility among teams. Equip your maintenance crew with checklists and training to perform quick, recurring assessments.

Continuous Improvement: Encourage a culture where feedback from frontline workers those who operate the equipment daily is integrated into maintenance schedules. This helps refine inspections and identify new risk areas that may have been previously overlooked.

Developing a Customized Preventative Maintenance Schedule

Asset Inventory and Prioritization:

Begin by cataloging every major piece of equipment and infrastructure component on the golf course.

Classify items by criticality equipment that directly affects play (e.g., mowers, tractors, irrigation controllers) should be prioritized over ancillary assets.

Frequency and Task Breakdown:

Daily and Weekly Tasks: Routine visual checks, cleaning, and minor adjustments are essential to catch problems early.

Monthly and Quarterly Tasks: More comprehensive servicing including calibration, lubrication, and deep diagnostics should occur according to usage intensity and manufacturer recommendations.

Annual Reviews: A yearly, in depth audit of all machinery and infrastructural elements can identify long term trends and inform capital expenditure planning.

Documentation and Follow Up:

Use digital maintenance management systems to log each preventative task performed.

Schedule follow ups based on the severity of identified issues to ensure that all repairs are addressed in a timely manner.

Real World Example

A superintendent in a high traffic resort course noted that after instituting a rigorous preventative maintenance plan, the number of unplanned equipment downtimes dropped by nearly 30% in one year. Their system involved not only strict adherence to scheduled tasks but also regular reviews of historical maintenance data. With each season, adjustments were made based on observed wear patterns and sensor analytics. This proactive approach not only saved money but also significantly enhanced the overall reliability of the fleet and infrastructure.

Diagram Idea

A useful visual might be a Gantt chart or a calendar layout that maps out daily, weekly, monthly, and annual maintenance cycles for different types of equipment and infrastructure. This chart would serve as both a planning tool and an at a glance reference to ensure nothing is overlooked throughout the year.

Checklist Preventative Maintenance Schedule Essentials

Develop a comprehensive inventory of all critical assets and infrastructure.

Create and implement a maintenance calendar with daily, weekly, monthly, and annual tasks.

Integrate digital tracking and remote sensor data into maintenance planning.

Document every maintenance activity and schedule follow ups as needed.

Regularly review historical maintenance records to refine the schedule.

Train staff on the importance and procedures of preventative maintenance.

Preventative maintenance is not just about avoiding breakdowns it is about continuously optimizing operations and ensuring that every investment in equipment and infrastructure delivers maximum value over its lifetime.

Conclusion: Integrating Maintenance into the Course's Lifeblood

Equipment and infrastructure maintenance is the unsung hero of successful golf course management. This chapter has taken a deep dive into the strategies, tools, and practices that support the core functions of maintaining quality equipment and infrastructure:

Routine Equipment Servicing ensures that every piece of machinery from mowers to aerators runs efficiently, safely, and reliably, minimizing downtime and prolonging asset life.

Cart Path and Clubhouse Landscaping enhance the visible appeal and functionality of the course environment, directly influencing the first impressions of visitors and members.

Preventative Maintenance Schedules form the backbone of a long term operational strategy, using technology and systematic documentation to preempt problems before they become disruptive.

By integrating these maintenance practices, superintendents can achieve a harmonious balance between operational efficiency, safety, and visitor satisfaction. A well maintained fleet and infrastructure not only keep the course running smoothly day to day but also safeguard the significant capital investments in equipment and facilities, ensuring the course remains a benchmark of excellence.

End of Chapter Checklist: Equipment & Infrastructure Maintenance

Routine Equipment Servicing:

Perform daily pre operation visual and functionality checks.

Schedule weekly cleaning, lubrication, and fluid level assessments.

Conduct monthly detailed diagnostics and calibrations.

Maintain comprehensive maintenance logs and follow manufacturer guidelines.

Cart Path and Clubhouse Landscaping:

Inspect cart paths monthly for damage and drainage issues.

Plan for seasonal resurfacing and proper sealing of surfaces.

Regularly update clubhouse landscaping with seasonal plantings and routine care.

Preventative Maintenance Schedules:

Establish an asset inventory and prioritize maintenance needs.

Create a detailed calendar outlining daily, weekly, monthly, and annual tasks.

Integrate remote monitoring and data analytics to refine maintenance schedules.

Document all maintenance activities and conduct periodic reviews.

Final Thoughts

Equipment and infrastructure maintenance is not merely a set of routine tasks it is the foundational practice that ensures the operational heartbeat of the golf course remains strong throughout every season. A robust maintenance program serves as a strategic investment in both the longevity of high value equipment and the physical appeal of the entire facility. By combining regular inspections, proactive servicing, and technology driven monitoring, superintendents can avoid unexpected breakdowns, mitigate costly emergency repairs, and ultimately safeguard the course's reputation for excellence.

The comprehensive approach presented in this chapter demonstrates that every component from the intricacies of daily equipment servicing to the periodic enhancements of cart paths and clubhouse landscaping plays a pivotal role in the overall guest experience. When equipment is maintained with precision, the turf remains pristine, the carts glide effortlessly, and the clubhouse serves as a welcoming front that reinforces the course's premium image. This level of stewardship not only protects capital investments but also creates a safer work environment, ensuring that the team can perform their tasks without the distraction of recurring malfunctions or safety hazards.

Moreover, establishing a preventative maintenance schedule is about building resilience. By anticipating potential issues through systematic, scheduled reviews and integrating smart sensors and digital record keeping, superintendents can transform reactive crises into planned and managed upgrades.

This data driven approach enables continuous improvement, whereby each servicing session informs future maintenance decisions. In essence, each task performed today contributes to a historical record that enhances operational planning and fuels cost savings over the long term.

Looking ahead, the dynamic nature of modern equipment and infrastructure demands that superintendents remain agile. The field is ever evolving, with new technologies and materials emerging that can further extend the lifecycle of assets and enhance efficiency. Adopting such innovations, while integrating them into traditional maintenance frameworks, provides a competitive edge that sets the course apart from its peers. It is this blend of time tested practices and forward thinking upgrades that fortifies a golf course against the unpredictable pressures of weather, usage, and market fluctuations.

Ultimately, the takeaway from this chapter is clear: consistent, detailed maintenance is the unsung hero behind smooth operations and an exceptional guest experience. It is the constant, quiet work behind the scenes that, when performed diligently, not only preserves but enhances the quality and longevity of every element on the course. By embracing both routine servicing and a preventative, data informed approach, golf course superintendents ensure that every tractor, cart path, and landscaped area remains in peak condition; thus, transforming maintenance from a mere obligation into a strategic asset that underpins the entire operational ecosystem.

Chapter 4: Spring Preparation & Early Season Recovery

Spring is a transformative season on any golf course. As winter recedes, turf grass emerges from dormancy with both hidden damage and new growth potential. This chapter explores the critical early steps required to assess turf health, revitalize the soil with aeration and over seeding, and accurately apply pre emergent herbicides to prepare the course for a vigorous growing season.

In spring, the superintendent's focus shifts from defensive measures to proactive rejuvenation. The goals are clear: accurately gauge the condition of the turf, restore compromised areas, and preemptively control weeds that can undermine turf recovery. Achieving these goals requires a well orchestrated plan that leverages both scientific protocols and experiential know how.

Assessing Turf Conditions After Winter Dormancy

Understanding Post Winter Turf Health

After long winter dormancy, the turf may show uneven recovery, compaction, and sometimes even disease symptoms triggered by low temperatures or excessive moisture. Accurate assessment is the critical first step in determining subsequent management actions. During this period, superintendents are tasked with "reading" the course like a diagnostic report, taking note of areas where the turf is slow to green up, regions showing discoloration, and patches of potential winter injury.

Fusarium Patch: Identification and Management

Fusarium patch is a persistent fungal disease that often becomes problematic on golf courses located in cool, wet climates particularly during the transitional periods of spring and fall. Caused by various Fusarium species, this disease can rapidly impair turf quality, leading to unsightly, water soaked lesions and areas of thinning turf. Early identification and a proactive management strategy are crucial for maintaining both the aesthetics and playability of a course.

Understanding Fusarium Patch

Symptoms and Visual Indicators

Discoloration and Water Soaked Appearance: Infected turf typically displays a grayish to yellowish hue, often accompanied by a water soaked or glistening appearance. These patches can spread quickly if conditions favor the fungus.

Rapid Development: After prolonged periods of moisture such as during extended spring showers signs of fusarium patch may appear within just a few days. In severe instances, large, confluent patches can form, undermining both turf density and vigor.

Localized Damage: While patches may initially be confined to smaller areas, they can merge over time. Areas with poor airflow, shading, or inadequate drainage are especially vulnerable.

Environmental Conditions Contributing to Fusarium Patch

Cool, Humid Climates: Fusarium thrives in temperatures ranging from 60°F to 75°F (15°C–24°C), particularly when coupled with high humidity. In regions where spring and fall are characterized by persistent wet weather, the risk of outbreak increases.

Prolonged Moisture: Extended periods of leaf wetness due, for example, to heavy rainfall or inefficient irrigation practices create the perfect setting for the fungus to establish itself.

Compromised Turf: Turf that is recovering from winter dormancy or that has been stressed by aggressive fertilization routines (especially those with high nitrogen levels) may be more susceptible to fusarium invasion.

Integrated Management Strategies

Cultural Practices

Optimized Irrigation: Adjust irrigation schedules to prevent prolonged periods of surface wetness. Techniques such as drip irrigation or using low volume spray systems can help minimize excess moisture on the turf leaves.

Improved Air Circulation: Regular aeration not only alleviates soil compaction but also increases air movement within the turf canopy. This can reduce humidity levels around the plant material and hinder the conditions that favor fungal proliferation.

Balanced Fertilization: Avoid excessive nitrogen applications in early spring. Instead, adopt a balanced fertilization strategy that promotes steady, resilient growth instead of lush, overly tender turf which is more prone to fungal infections.

Proper Traffic Management: Limit heavy foot and cart traffic on vulnerable areas during periods of high moisture. Reducing stress on the turf can help maintain its natural resistance to disease.

Chemical Controls

Fungicide Applications: In situations where environmental conditions and early symptoms suggest a high risk of fusarium patch, timely fungicide treatments can be critical. Look for fungicides with efficacy against Fusarium species and always follow the manufacturer's guidelines on dosage and application timing.

Integrated Approach: Chemical treatments are most effective when combined with sound cultural practices. For example, a fungicide application immediately after a rainfall event paired with adjusted irrigation can help suppress an emerging outbreak.

Rotation of Products: To prevent the development of fungicide resistance, rotate between different classes of fungicides as recommended by industry guidelines.

Monitoring and Early Intervention

Routine Inspections: Conduct frequent, thorough inspections of high risk areas especially shaded regions and areas with poor drainage. Early detection is key; catching symptoms before the disease spreads allows for targeted interventions.

Documentation and Data Analysis: Maintain detailed records of fusarium incidence, including the timing, extent of affected areas, and any environmental conditions associated with outbreaks. This historical data can help refine future management strategies and predict periods of high risk.

Use of Technology: Consider integrating remote monitoring tools, such as moisture sensors and weather tracking systems, to better predict the onset of conditions favorable for fusarium patch. This proactive approach can guide the timing of both cultural and chemical interventions.

Diagram Concept

Imagine a flowchart that illustrates the lifecycle of fusarium patch:

Environmental Conditions (cool temperatures, prolonged moisture, high humidity)

Initial Infection (localized water soaked lesions appear)

Disease Progression (patches expand and merge)

Intervention Points (adjust irrigation, apply fungicides, improve aeration)

Recovery and Monitoring

This diagram can serve as an at a glance guide for identifying key intervention points that are critical for disease management.

Checklist Fusarium Patch Management Essentials

Early Detection: Regularly inspect the turf, focusing on shaded and low drift areas, for signs of discoloration and water soaked lesions.

Optimize Irrigation: Adjust practices to minimize prolonged moisture on the turf surface.

Enhance Airflow: Implement regular aeration to improve drainage and reduce humidity around the turf.

Balanced Fertilization: Use a tailored fertilization approach that avoids overly lush growth vulnerable to disease.

Fungicide Strategy: If early symptoms appear, apply recommended fungicides in a timely and rotated manner.

Record Keeping: Document incidences and correlate with weather events to fine tune future fusarium prevention tactics.

Final Thoughts on Fusarium Patch

Fusarium patch presents a significant challenge, particularly for courses in cool, wet climates.

However, with vigilant monitoring, integrated cultural practices, and targeted fungicide applications, superintendents can effectively mitigate its impact. By understanding the environmental triggers and deploying a proactive management strategy, you can stave off the rapid spread of this disease ensuring that your turf not only recovers from winter dormancy but thrives throughout the season.

Placing strong emphasis on fusarium patch management within your spring preparation protocols transforms potential setbacks into manageable challenges. In doing so, you safeguard your course's aesthetic quality and operational performance, allowing your turf to serve as a resilient foundation for the entire playing season.

This standalone section can be integrated as a distinct heading within Chapter 4. Feel free to adjust its placement or content as needed to seamlessly merge with your overall chapter structure. Would you like any further modifications or additional details on this section?

Additional Turf grass Diseases: Identification and Management

In addition to fusarium patch, golf courses in many regions face challenges from several other turf grass diseases. These diseases can compromise the aesthetic quality, playing conditions, and long term health of the turf if not managed promptly. The most common ones include Dollar Spot, Brown Patch, and Pythium Blight. Each has unique symptoms, thrives under specific environmental conditions, and requires targeted cultural and chemical interventions.

Dollar Spot

Overview and Symptoms

Dollar Spot is one of the most frequently encountered turf grass diseases on golf courses, particularly in warm, humid conditions. It is caused by a fungal pathogen that produces small, circular, silver dollar sized lesions on the turf blades. Early symptoms typically include pinpoint yellow or straw colored areas that quickly develop into more defined patches. Over time, these patches may merge, leading to significant thinning and uneven turf density.

Environmental Conditions

Warm and Humid Weather: Dollar Spot thrives when temperatures are moderately warm (70°F–80°F) and humidity is high.

Over fertilization Risks: Excessive nitrogen can promote lush, tender growth that is more susceptible to the disease.

Poor Air Circulation: Areas with inadequate airflow or excessive thatch buildup often see higher disease incidence.

Management Strategies

Cultural Practices:

Balanced Fertilization: Use moderate nitrogen levels and slow release fertilizers to avoid overly lush growth.

Proper Mowing: Maintain appropriate mowing heights to discourage dense thatch and improve air circulation.

Improved Irrigation: Water early and in a controlled manner to prevent prolonged leaf wetness.

Chemical Treatments:

Fungicide Applications: Select fungicides labeled for Dollar Spot control. Apply preventatively during high risk periods or at the first sign of symptoms.

Rotation of Fungicides: To prevent resistance, rotate between different fungicide classes as per regional recommendations.

Checklist Dollar Spot Management

Monitor turf for small, circular patches of yellow or straw coloration.

Apply cultural practices to maintain balanced growth and reduce thatch.

Time fungicide treatments during periods of high humidity and warm temperatures.

Rotate fungicide products to avoid resistance buildup.

Brown Patch

Overview and Symptoms

Brown Patch, predominantly affecting cool season grasses like bentgrass and perennial ryegrass, manifests as circular, brown patches with a darker, wet colored border. These patches often appear rapidly on greens and heavily visited fairways during periods of warm, humid days followed by cool nights. The disease can cause significant turf loss in localized areas and may result in an uneven playing surface if not managed in time.

Environmental Conditions

Cool Nights and Warm Days: Brown Patch is typically favored by a combination of warm daytime temperatures (65°F–75°F) and cooler nights.

High Humidity and Moisture: Prolonged dew or excessive irrigation can maintain the necessary moisture for fungal growth.

Stressed Turf: Turf that is weakened by physiological stress or over fertilization is particularly vulnerable to infection.

Management Strategies

Cultural Practices:

Optimized Irrigation: Limit water applications to prevent extended periods of wetness, especially in the evening.

Reduced Fertilization Rates: Avoid overloading the turf with nitrogen during peak risk periods to encourage stronger, less susceptible growth.

Enhanced Air Circulation: Aerate and dethatch to reduce dense thatch layers and increase airflow, lowering humidity around the turf.

Chemical Treatments:

Fungicide Interventions: Use fungicides specifically effective against Brown Patch as soon as symptoms are detected. Follow label directions and consider using a preventive spray during high risk weather conditions.

Checklist Brown Patch Management

Inspect turf regularly for circular brown patches with dark borders.

Adjust irrigation schedules to minimize evening moisture.

Moderate nitrogen applications during periods of risk.

Implement aeration and dethatching where necessary.

Initiate fungicide treatment promptly at symptom onset

Pythium Blight

Overview and Symptoms

Pythium Blight is a rapidly spreading disease most common in extremely humid conditions or areas with poor drainage. It is characterized by greasy, water soaked lesions on the turf blades that quickly transition to a slimy, cotton like growth on the affected areas. This disease can devastate turf stands in a short period if favorable conditions persist and is often more aggressive on stressed or newly overseeded areas.

Environmental Conditions

Excessive Moisture: Pythium Blight thrives in situations where there is prolonged leaf wetness due to heavy rainfall or inefficient drainage.

High Humidity: The disease is promoted by humid environments and can be exacerbated in microclimates where moisture lingers.

Stressed Turf: Areas experiencing rapid growth or stress (after over seeding, for instance) are particularly vulnerable.

Management Strategies

Cultural Practices:

Improved Drainage: Ensure that drainage systems are functioning properly to prevent water pooling.

Irrigation Adjustments: Reduce the frequency and volume of water applications during high risk periods. Water early in the day to allow the turf to dry during warmer hours.

Stress Reduction: Avoid over seeding or aggressive fertilization practices that can make turf more susceptible to Pythium.

Chemical Treatments:

Fungicide Options: Use fungicides targeted against Pythium Blight, following careful application timing ideally as soon as initial symptoms appear.

Integration with Cultural Practices: Combine fungicide applications with proactive cultural adjustments to sustainably reduce disease pressure.

Checklist Pythium Blight Management

Maintain well functioning drainage to prevent prolonged leaf wetness.

Adjust irrigation practices to reduce excessive moisture, especially during humid periods.

Monitor for signs of greasy, water soaked lesions developing into cottony patches.

Initiate fungicide treatments promptly when symptoms are detected.

Avoid practices that overstimulate rapid turf growth during high risk periods.

Diagram Concept

Imagine a composite visual that maps out the common turf grass diseases. The diagram could include branches for Dollar Spot, Brown Patch, and Pythium Blight, each featuring key symptoms, optimal environmental conditions, and pinpointed management strategies. This diagram serves as an at a glance reference for quick disease identification and subsequent action steps.

Final Thoughts on Additional Turf grass Diseases

Proactive management of turf grass diseases beyond fusarium patch is essential for maintaining a consistent, high quality golf course. Dollar Spot, Brown Patch, and Pythium Blight all present unique challenges that, if left unmanaged, can lead to diminished turf density, uneven playing surfaces, and increased maintenance costs. By integrating sound cultural practices with timely fungicide applications and by leveraging careful disease monitoring a superintendent can mitigate these risks effectively.

This additional section is designed to empower you with the knowledge and actions required to combat these common turf diseases. With a comprehensive approach that combines regular inspection, tailored cultural practices, and targeted chemical interventions, your turf can remain resilient and pristine, delivering both exceptional playability and aesthetic appeal throughout the season.

Checklist Overall Additional Disease Management

Maintain a routine turf inspection schedule to detect early disease symptoms.

Adjust cultural practices (irrigation, fertilization, aeration) to create an environment less conducive to disease proliferation.

Apply fungicides as needed, while carefully rotating products to avoid resistance.

Document disease occurrences to refine future management protocols.

Key Steps in Turf Assessment

A. Visual and Physical Evaluations

Color and Density Observations: Start with a systematic walk through of the course. Look for variations in color vivid green regions indicate healthy growth, while darker or yellow patches may suggest nutrient deficiencies, excess moisture, or disease. Take note of turf density and uniformity.

Texture and Compaction Checks: Gently probe the turf. In some areas, the soil may have become compacted due to freeze thaw cycles or heavy winter traffic, limiting root expansion and water penetration. Use a soil probe or a simple screwdriver to evaluate resistance.

B. Diagnostic Testing and Sampling

Turf Growth Measurements: Document the height of turf blades in various sections (greens, fairways, tees, and roughs). This provides baseline data for future mowing and fertilization plans.

Soil Moisture and Temperature Readings: Utilize portable moisture meters and thermometers to gauge soil conditions. Optimal conditions for spring recovery include a balance of adequate moisture and encouraging temperature gradients that stimulate root activity.

Damage Mapping: Create a field map that highlights "trouble spots" across the course. Mark areas with visible symptoms of winter injury, such as turf thinning or discoloration, to direct future interventions.

Real World Example: A superintendent in New England discovered that after an unusually wet winter, certain fairway zones exhibited significant compaction and waterlogging. By systematically assessing the turf and mapping these zones, the team was able to apply targeted aeration and drainage improvements, which improved overall recovery.

Diagram Idea: A seasonal timeline or visual dashboard that charts key turf indicators (color, density, moisture, and temperature) can help superintendents visualize how recovery progresses over the first few weeks of spring. This "turf health map" can be revisited regularly to track improvements.

Checklist Turf Condition Assessment

Walk the course systematically, noting color and density variations.

Use a soil probe to check for compaction in key areas.

Measure turf blade height and record differences among various zones.

Utilize moisture and temperature meters for precise readings.

Create and update a damage map highlighting areas needing intervention.

By rigorously assessing turf conditions immediately after winter, superintendents can make informed decisions on the types and intensities of interventions required.

Aeration and over seeding Strategies

Restoring Soil Structure and Enhancing Turf Density

Aeration and over seeding are powerful tools for restoring a golf course after winter. Aeration relieves soil compaction, improves drainage and oxygen flow to the roots, and creates an ideal seedbed for over seeding. over seeding, in turn, rejuvenates thin or damaged turf by introducing new grass cultivars that are more resilient or better suited to prevailing conditions.

Aeration: Techniques, Timing, and Tools

A. Aeration Methods: Spike vs. Plug Aeration

Spike Aeration: Involves puncturing the soil with solid tines. While this method creates channels for air and water, it may not be as effective in breaking up compacted soil compared to plug aeration.

Plug (Core) Aeration: Removes small cores of soil from the turf, reducing compaction more dramatically and allowing for better infiltration of water, nutrients, and overseeded material. This method is generally favored for high use areas that have suffered severe compaction over winter.

B. Timing and Frequency

Pre over seeding Timing: Ideally, aeration should be performed as early as possible in the spring to coincide with the turf's reactivation and before over seeding. This timing creates a receptive environment for seeds to lodge into the soil.

Repetition Based on Need: In severely compacted areas, double aeration (aerating, then over seeding followed by another pass) may be warranted. However, caution must be exercised to not overly disturb the turf.

C. Equipment and Best Practices

Modern Aeration Machines: Use equipment that minimizes turf disruption. Newer models are designed to extract uniform plugs without leaving excessive holes, which can hasten recovery.

Post Aeration Management: Following aeration, lightly rake the surface to redistribute soil plugs and prepare a smooth seedbed for over seeding.

Real World Example: At a course in the southeast, the superintendent switched from spike aeration to plug aeration in problematic fairways. This change improved seed to soil contact during over seeding, resulting in a 30% increase in turf density over the following season.

Over seeding Strategies

A. Choosing the Right Seed Mix

Tailored Seed Blends: Use seed varieties that complement the existing turf while resisting common diseases and wear. Select varieties that germinate quickly and establish strong root systems.

Seeding Rate: Follow recommended over seeding rates, as too much seed can lead to overcrowding, while too little may not adequately fill in damaged areas.

B. Over seeding Techniques

Even Distribution: Use high quality spreaders to ensure even distribution of seed across aerated areas. Uniform seeding is critical to avoid patchy turf recovery.

Incorporating Soil Amendments: Mix in organic matter or starter fertilizers that boost seed germination and early growth. These amendments also provide a protective cushion that encourages root establishment.

Post Seeding Care: After over seeding, maintain light irrigation to keep the seedbed consistently moist and encourage successful germination. Avoid excessive traffic until the new turf is established.

Diagram Idea: A dual phase diagram can illustrate both the aeration process and over seeding. For instance, the first phase shows how plug aeration creates holes where new seeds can be deposited, and the second phase displays the emergence of vigorous, dense turf as a result of this combined strategy.

Checklist Aeration and over seeding Essentials

Decide on the appropriate aeration method (plug aeration is generally preferred for deep compaction).

Schedule aeration to align with early spring and before over seeding.

Use up to date aeration machinery to achieve uniform soil core removal.

Select a seed mix that complements current turf conditions and local environmental factors.

Oversee even seed distribution through calibrated equipment.

Follow aeration and over seeding with proper post treatment irrigation and soil amendment application.

Through careful aeration and over seeding, superintendents can transform stressed, compacted turf into a vibrant, resilient playing surface primed for an active season of play.

Pre emergent Herbicide Applications

Preventing Weeds Before They Emerge

Weeds are opportunistic invaders that quickly take advantage of the early spring conditions when turf grass is recovering. Pre emergent herbicides are a key tool for superintendents, as they create a chemical barrier that prevents weed seeds from germinating. These applications protect the turf, ensuring that the energy of the beginning of season growth is directed toward healthy grass rather than competing weeds.

Timing and Selection of Herbicides

A. Choosing the Right Pre emergent Herbicides

Herbicide Types: Select herbicides specifically designed for turf grass systems. There are formulations targeting broadleaf weeds, grassy weeds, or a combination thereof.

Product Compatibility: Ensure that the chosen herbicide is compatible with the grass type and over seeding plans. Some products might inhibit seed germination if not applied correctly; hence, timing is critical.

B. Application Timing and Frequency

Early Spring Application: The application should occur early in the season ideally immediately after winter dormancy and before the majority of weed seeds have the chance to germinate.

Weather Considerations: Consider weather forecasts; applying herbicides during calm, dry periods helps prevent drift and ensures that the product adheres well to the target zones.

Integration with Other Practices: Coordinate pre emergent applications with aeration and over seeding schedules. For example, wait for aeration to settle before applying herbicides to ensure their effectiveness without disrupting seed germination in overseeded areas.

Techniques and Best Practices

A. Calibration and Uniform Application

Equipment Calibration: Use calibrated sprayers to achieve uniform distribution. Over application can injure the turf, while under application may allow weeds to break through.

Safety Measures: Follow all label recommendations for personal protective equipment (PPE) and safe handling. Ensure that staff – especially those handling chemicals – are properly trained and that environmental protocols are in place to prevent runoff into grassy areas beyond the target zones.

B. Post Application Care

Watering Guidelines: Some pre emergent's require a light irrigation (if conditions permit) to activate the chemical without causing runoff. Follow product specific guidelines to optimize effectiveness.

Monitoring Turf Response: After application, observe the turf closely. Look for any signs of unintended damage or stress, and adjust follow up maintenance practices (such as irrigation or fertilization) accordingly.

Real World Example: A superintendent in a coastal climate found that early pre emergent applications, combined with post treatment light irrigation, reduced broadleaf weed pressure by nearly 50% compared to previous years. This improvement not only contributed to a cleaner appearance but also allowed the overseeded turf to establish more uniformly, free from the competitive pressure of invasive weeds.

Diagram Idea

Consider a flow diagram that illustrates the pre emergent application process: starting with the timing decision based on soil temperature and weather, followed by equipment calibration, application procedures, and post application irrigation recommendations. This offers an at a glance guide for both planning and verification.

Checklist Pre emergent Herbicide Application Essentials

Select herbicides that are compatible with the turf type and over seeding strategy.

Plan applications for early spring, before the majority of weeds emerge.

Calibrate sprayers to ensure uniform and accurate chemical distribution.

Observe all safety protocols and PPE guidelines during chemical handling.

Follow post application watering guidelines to activate products without causing runoff.

Monitor turf response and adjust cultural practices as needed.

Pre emergent treatments, when timed and executed correctly, form a critical line of defense against invasive weeds. Blocking weed emergence preserves the vitality of the recovering turf, ensuring that the nutrients, water, and light required for a robust start are not siphoned away by unwanted competitors.

Conclusion

Spring preparation and early season recovery are the cornerstones of a thriving golf course. In this chapter, we explored the critical tasks that set the stage for a successful growing season:

Assessing Turf Conditions After Winter Dormancy: Conducting thorough visual and physical assessments ensures that you identify areas of potential trouble early on. The use of diagnostic tools and mapping techniques can lead to informed decisions that drive seasonal strategies.

Aeration and over seeding Strategies: By relieving soil compaction and rejuvenating thin or damaged areas, aeration and over seeding work hand in hand to rebuild turf density. Employing the correct techniques, timing interventions appropriately, and integrating post treatment care maximize the benefits of these practices.

Pre emergent Herbicide Applications: Preventative chemical barriers are vital for keeping weeds from colonizing vulnerable turf in the early spring. Meticulous application guided by calibrated equipment and product specific protocols ensures that your turf's emerging growth remains uninterrupted by invasive species.

Together, these practices create a harmonious, proactive approach to spring turf management. They not only restore the turf's aesthetic and functional quality but also set in motion a cycle of regeneration and sustained health that will benefit the entire course throughout the season.

By integrating rigorous assessment procedures with targeted rejuvenation strategies and preventative measures, superintendents can confidently transition the course from the dormancy of winter into a vibrant, growing landscape ready for peak play. This early investment in proper spring preparation pays dividends in turf resilience, improved playability, and long term cost efficiency.

End of Chapter Checklist: Spring Preparation & Early Season Recovery

Turf Assessment:

Conduct a systematic visual and physical inspection.

Record turf density, color, and blade height.

Map areas of compaction and winter injury.

Aeration and over seeding:

Choose the appropriate aeration method for each zone.

Schedule aeration and over seeding aligned with turf recovery.

Select seed blends that support turf regeneration.

Implement post treatment irrigation and amendments.

Pre emergent Herbicide Application:

Select herbicides compatible with both turf and over seeding plans.

Calibrate application equipment for uniformity.

Follow proper timing protocols, considering weather and soil conditions.

Monitor and adjust post application care based on turf response.

Final Thoughts

The early spring period is a time of renewal and opportunity a season when proactive management can redefine the trajectory of turf performance for the entire year. By meticulously assessing turf conditions, implementing precise aeration and over seeding techniques, and applying pre emergent herbicides at optimal times, superintendents transform their courses into vibrant, resilient landscapes poised for robust growth.

Embracing these practices not only supports immediate turf recovery but also builds the foundation for sustained excellence. As the turf begins its transformation from winter dormancy to energetic green waves, every well executed task magnifies the course's potential. Superintendents who invest deeply in these early season strategies will notice not only improved turf health and uniformity but also a more satisfying playing experience for golfers throughout the season.

This chapter's detailed strategies, supported by visual aids and actionable checklists, serve as a critical guide for ensuring that the transition from winter to spring is managed with precision and foresight. In the dynamic world of golf course management, early interventions make all the difference. With a clear, systematic approach to spring preparation and early season recovery, you set the stage for a season marked by vigorous turf growth, consistent performance, and operational success.

Chapter 5 Summer Maintenance & Heat Management

Summer brings a host of challenges that require precision, planning, and proactive management by golf course superintendents. During the hottest months of the year, increasing evaporation rates, fluctuating moisture levels, and the intensification of pest and disease pressures create an environment where every decision impacts turf health and course playability. In this chapter, we explore three critical areas of summer maintenance and heat management:

1. Managing Turf Hydration Levels and Drought Stress

2. Pest and Disease Control Strategies

3. Heat Stress Prevention Techniques

Each section is presented in depth, with practical strategies, process explanations, diagram ideas, and detailed checklists. The goal is to arm you with the knowledge necessary to maintain resilient turf, reduce water waste, control pest populations, and mitigate the damaging effects of extreme heat.

Introduction

As the weather heats up, turf grass enters a phase where rapid water loss and high temperatures challenge every maintenance strategy. Hydration levels become critical not only to keep the turf green but also to ensure that the root system remains vigorous under stress. Meanwhile, with summer comes an increased pressure from both insect pests and fungal pathogens that exploit the weakened, overheated turf. Additionally, heat stress can lead to injury, discoloration, and an overall decline in turf quality if not managed properly.

This chapter provides a systematic approach to summer management. We begin by exploring strategies to manage turf hydration and alleviate drought stress. Next, we address the twin challenges of pest and disease control during the sweltering months, where even small infestations or outbreaks can quickly escalate. Finally, we focus on proactive methods to prevent heat stress, ensuring that turf remains robust even when temperatures soar.

By mastering these aspects, you will not only increase the resilience of your course but also optimize water usage, reduce maintenance costs, and create a superior playing environment that endures the rigors of summer.

Managing Turf Hydration Levels and Drought Stress

Understanding Turf Hydration in Summer

In the heat of the summer, the balance between adequate hydration and over irrigation is delicate. The combination of high temperatures, intense sunlight, and low humidity escalates evapotranspiration rates dramatically. This creates a scenario where water may be rapidly lost before it ever reaches the root zone. Drought stress sets in when the turf does not receive the water it needs to sustain cell function, leading to wilting, discoloration, and, ultimately, compromised playability.

Key Challenges and Indicators

Evapotranspiration (ET) Increase: Under high heat, ET rates spike. Understanding local ET rates is critical because watering schedules must be adjusted based on real time weather data and seasonal trends.

Soil Moisture Deficits: Frequent measurements using soil moisture sensors reveal how deep the water penetrates. Shallow watering encourages superficial rooting, making turf more vulnerable to drought.

Visible Symptoms: Yellowing, wilting, or a bluish hue in turf can be early signs of drought stress. Monitoring these symptoms regularly on high use areas (e.g., fairways, tees) is crucial for timely intervention.

Strategies for Optimizing Hydration

A. Precision Irrigation Scheduling

Smart Controllers: Upgrade to irrigation controllers that integrate local weather data. These systems adjust watering schedules automatically based on factors like temperature, humidity, and wind speed.

Zonal Irrigation: Divide the course into zones so that areas with different sun exposures or soil types can be watered independently. This allows for fine tuning water application to match the exact needs of different parts of the course.

Deep, Infrequent Watering: Focus on delivering water deeply to encourage robust root growth. Deep watering practices preferably early in the morning help the turf develop a deeper root network, increasing its ability to access water stored deeper in the soil.

B. Water Conservation Techniques

Soil Moisture Sensors: Utilize soil moisture sensors to provide real time feedback. These devices can trigger irrigation only when necessary, reducing water waste during periods when the turf already has adequate moisture.

Surfactants and Wetting Agents: Apply wetting agents periodically to ensure that water penetrates hydrophobic (water repellent) soils. This is particularly useful in sandy soils, where water infiltration can be slow.

Mulching and Organic Amendments: Incorporate organic matter into the soil to improve its water holding capacity. Organic amendments improve soil structure and increase the water retention ability in erosion prone or sandy areas.

C. Monitoring and Adjustments

Regular Field Walks and Data Reviews: Combine daily visual inspections with sensor data. Look for dry patches, signs of stress, and uneven turf color. Compare these observations with irrigation system output data to adjust schedules promptly.

Budgeting Water Use: Create an irrigation budget for the high summer months. This involves calculating the anticipated ET and adjusting water allocation accordingly ensuring that water is used efficiently without compromising turf health.

Real World Example

At a resort course in a semi arid region, the superintendent installed a network of digital soil moisture sensors along with smart irrigation controllers. Without these implementations, the course was experiencing significant water loss. The new system allowed them to reduce irrigation by 20% while maintaining a lush, green playing surface. They noticed an extended rooting depth, which made the turf more resilient during subsequent dry spells.

Diagram Idea

Imagine a flow diagram that shows the irrigation decision process for summer:

1. Start with real time soil moisture data.

2. Compare against target moisture levels.

3. Adjust watering duration or frequency accordingly through smart controller inputs.

4. Final output: Optimized water delivery per zone.

Checklist Managing Turf Hydration and Drought Stress

Install and calibrate soil moisture sensors in key areas.

Program smart irrigation controllers to adjust for real time weather conditions.

Divide the course into irrigation zones based on soil type and exposure.

Schedule deep watering sessions early in the morning.

Apply wetting agents periodically to enhance infiltration.

Incorporate organic amendments to boost soil water retention.

Create and review a water usage and ET budget for the summer months.

Perform regular visual inspections and compare with sensor data.

By integrating these precision irrigation and conservation techniques, you can manage turf hydration more effectively and prevent drought related deterioration, setting a robust foundation for summer turf vigor.

Pest and Disease Control Strategies

The Summer Pest and Pathogen Challenge

Summer is not only tough on turf because of heat but also because it creates an ideal environment for pests and diseases. Warmer temperatures and higher humidity levels can accelerate the life cycles of many insects and pathogens, leading to outbreaks that can quickly spread across the course. Effective summer management requires an integrated approach to pest and disease control that merges cultural practices, biological controls, and, when necessary, chemical treatments.

Common Summer Pests and Diseases

A. Insect Pests

Grubs and Soil Dwelling Larvae:

Threats: These pests feed on the roots and can cause significant thinning of turf if large populations persist.

Monitoring: Regular soil sampling and visual inspections help detect early infestations.

Control: Use biological controls (e.g., beneficial nematodes) or targeted insecticides when threshold levels are exceeded.

Chinch Bugs:

Threats: Chinch bugs damage turf by feeding on the sap, which can create brown, dead patches.

Monitoring: Look for symptomatic areas on fairways and greens especially in high turf density areas.

Control: Employ insecticide sprays judiciously while integrating cultural strategies like proper mowing and irrigation management to discourage pest habitation.

Sod Webworms and Billbugs:

Threats: these caterpillar like insects feed on the leaf blades and stems, reducing turf quality.

Monitoring: Use pheromone traps or conduct scouting walks during peak activity periods.

Control: Apply appropriate biological or chemical controls based on established Integrated Pest Management (IPM) protocols.

B. Fungal Diseases

Summer Patch (Anthracnose):

Symptoms: Circular, tan patches that expand rapidly, often seen on stressed turf under high heat.

Management: Optimize mowing, fertilization, and watering practices to reduce turf stress; utilize fungicides if outbreaks are severe.

Powdery Mildew and Other Foliar Diseases:

Symptoms: A white, powdery fungus that can form on the leaves during hot, dry spells or late summer, affecting both aesthetics and turf health.

Management: Increase air circulation with proper mowing and avoid excessive nitrogen applications that can predispose turf to disease.

Integrated Pest Management (IPM)

IPM is a systematic approach combining multiple strategies that minimize pest damage while reducing reliance on chemical treatments. Key components include:

Cultural Controls

Mowing Practices:

Maintain optimal mowing heights; cutting too short can stress turf and make it more vulnerable to pests.

Vary mowing patterns to reduce pest habitat development and avoid creating a uniform environment that pests can easily exploit.

Fertilization Management:

Provide balanced nutrition to keep turf healthy. Over fertilization, particularly with nitrogen, can lead to overly lush, succulent growth that is more attractive to pests.

Use slow release fertilizers that minimize sudden nutrient surges which could trigger pest outbreaks.

Irrigation Practices:

Ensure proper irrigation to prevent both water stress and overly wet conditions that favor fungal growth.

Schedule irrigation to reduce leaf wetness at times that coincide with peak pest activity.

Cultural Clean Up:

Remove thatch and clippings regularly to eliminate breeding grounds for pests.

Use raking and aeration to prevent the buildup of organic debris, which can harbor disease spores.

Biological Controls

Beneficial Insects and Microorganisms:

Introduce predatory insects (such as parasitic wasps) and beneficial nematodes to help naturally reduce pest populations.

Encourage a diverse microbial ecosystem in the soil through the use of organic amendments and minimal pesticide use.

Microbial Inoculants:

Some biological treatments involve applying microbial inoculants that outcompete pathogenic fungi by establishing beneficial populations in the turf environment.

Chemical Controls

Targeted Insecticides and Fungicides:

Apply chemical treatments only when necessary, based on scouting results and economic threshold levels.

Rotate chemical classes to prevent resistance and follow all label instructions to protect non target organisms.

Preventive vs. Curative Applications:

Employ preventive fungicide treatments during periods when environmental conditions favor disease, while reserving curative treatments for actual outbreaks.

Use spot treatments for localized insect infestations rather than whole course applications.

Monitoring and Decision Making

Regular Scouting:

Establish a routine inspection schedule daily during peak periods, weekly otherwise to detect early signs of pest or disease problems.

Use tools such as sticky traps, pheromone traps, and visual surveys to monitor insect populations and disease symptoms.

Data Driven Decisions:

Maintain records of pest counts, disease incidences, and weather patterns. This historical data can help predict outbreaks and guide timely interventions.

Collaborate with local extension services or turf specialists for guidance on threshold levels and treatment recommendations.

Diagram Idea

Visualize an integrated pest management cycle a circular diagram that begins with scouting, followed by cultural interventions, then the application of biological controls, and, as needed, targeted chemical treatments. Arrows indicate how each step informs the next, with regular monitoring loops that feed back into the decision making process.

Checklist Pest and Disease Control Essentials

Conduct daily or weekly pest and disease scouting across all turf areas.

Adjust mowing and irrigation practices to minimize pest friendly conditions.

Keep detailed records on pest levels, disease occurrences, and corresponding weather conditions.

Implement biological control measures where possible (beneficial insects, microbial inoculants).

Apply chemical controls only when scouting and threshold levels indicate a need.

Rotate fungicide and insecticide products to prevent resistance.

By systematically integrating cultural, biological, and chemical controls into your summer maintenance program, you can curb the threats posed by pests and diseases and maintain a healthier, more resilient turf environment.

Heat Stress Prevention Techniques

Understanding Heat Stress in Turf grass

High summer temperatures can directly harm turf grass by causing cellular damage, reducing photosynthesis, and disrupting water uptake. Heat stress is particularly problematic when coupled with drought, as limited water exacerbates the inability of turf to cool itself through transpiration. Effective heat stress prevention focuses on maintaining soil moisture, adjusting cultural practices, and sometimes even employing technological aids to reduce the ambient heat load on the turf.

Strategies to Prevent Heat Stress

A. Adjusting Irrigation Practices

Early Morning Watering:

Schedule irrigation early in the morning. This minimizes evaporation losses and provides the turf with a moisture reservoir that it can use throughout the day.

Early watering also allows leaves to dry before the heat of midday, reducing the risk of disease development.

Deep Rooting Initiatives:

As discussed in the hydration section, deep, infrequent watering encourages deep root growth. Roots that extend deeper into the soil can access more moisture and are less likely to succumb to surface heat stress.

Use of Soil Moisture Sensors and Smart Controllers:

Integrate technology that monitors soil moisture and adjusts watering schedules accordingly. Data driven irrigation ensures that the turf receives water only when needed, minimizing waste and reducing the chance of waterlogging.

B. Cultural Practices to Reduce Heat Load

Mowing Practices:

Raise mowing heights slightly in the peak of summer. Taller grass can provide shade to the soil, reducing evaporation and protecting the root zone from intense sunlight.

Use sharp mower blades to create clean cuts dull blades can cause ragged edges, making turf more vulnerable to heat injury.

Mulching and Organic Matter Applications:

Leave clippings on the ground when possible to act as a natural mulch. This organic cover helps regulate soil temperature and conserves moisture.

Incorporate compost or other organic matter into the soil to improve its water holding capacity and insulation.

Wetting Agents and Anti Transpirants:

Use wetting agents to ensure even water distribution and to increase infiltration within the root zone.

In some cases, anti transpirant products (which reduce water loss from the plant) can be applied to help turf conserve moisture, though these should be used judiciously and in conjunction with proper irrigation practices.

C. Secondary Strategies and Technological Aids

Shade Management:

Planting trees or installing shade structures near high traffic areas (like tee boxes and greens) can reduce overall surface temperatures. While not feasible on every course, small scale adjustments can create microclimates that favor turf survival.

Use of Reflective Mulches or Ground Covers:

In some strategic areas, reflective mulches can help deflect excessive sunlight and lower soil temperatures.

Proactive Nutrient Management:

Avoid high nitrogen applications that can promote tender, susceptible growth. Instead, use balanced or slow release fertilizers that encourage steady, resilient growth.

Monitoring for Heat Stress Symptoms

Visual Indicators:

Look for signs of wilting, discoloration, or scorched leaf margins. Early detection is critical to re adjusting management practices before irreversible damage occurs.

Measuring Soil Temperature:

Use soil thermometers to monitor temperature spikes in key areas. Ground temperatures above 90°F (32°C) can be especially dangerous for turf.

Plant Vigor Assessments:

Regularly assess the turf for resilience. Areas that recover quickly after irrigation are likely managing heat stress better than those that remain wilted or discolored.

Diagram Idea

A multi layered diagram could illustrate the impact of high temperatures on turf grass, showing layers of the soil profile with root zones, coupled with icons indicating irrigation times (e.g., early morning watering), mowing height adjustments, and areas where wetting agents are applied. Such a diagram would serve as a visual checklist to help superintendents quickly identify areas of concern.

Checklist Heat Stress Prevention Essentials

Water deeply in the early morning rather than in the afternoon.

Use smart irrigation controllers and soil moisture sensors to adjust watering frequency.

Raise mowing heights during peak summer months to provide additional shade to the soil.

Maintain mower blades and consider mulching to help regulate microclimates.

Apply wetting agents to ensure even water distribution.

Monitor turf visual symptoms and soil temperatures regularly.

Adjust nutrient management practices to avoid overly tender growth vulnerable to heat stress.

Consider additional protective measures such as reflective mulches or temporary shading in critical areas.

By combining these techniques, superintendents can create a multi faceted heat stress prevention program that not only conserves water but also supports lawn health under extreme summer conditions.

Conclusion

Summer maintenance is a multifaceted challenge that demands both strategic planning and precise execution. During this season, the turf requires a tailored approach that encompasses careful management of hydration levels, diligent control of pests and diseases, and proactive measures to prevent heat stress.

Managing Hydration and Drought Stress

Deep, calibrated irrigation strategies ensure that water reaches deeper into the soil, fostering resilient, deep rooted turf capable of withstanding dry spells. Technology through the use of smart controllers and soil sensors plays a vital role in adapting irrigation practices to real time conditions. Combining these practices with organic amendments and proper scheduling forms the backbone of successful water management in the heat.

Controlling Pests and Diseases

Integrated Pest Management (IPM) proves indispensable in summer maintenance. By focusing on cultural practices such as optimized mowing, balanced fertilization, and proper irrigation, superintendents can reduce pest populations and proactively address disease pressures. Regular scouting, combined with a strategic mix of biological and chemical controls, ensures that pests like grubs, chinch bugs, and fungal pathogens do not overpower the turf.

Preventing Heat Stress

Heat stress, which can rapidly degrade turf quality, is addressed by adjusting irrigation schedules, modifying cultural practices, and adopting technological aids. Early morning watering, increased mowing heights, and the use of wetting agents or anti transpirants help safeguard the turf's health during extreme conditions. Proactive monitoring both visual and via soil temperature data allows timely interventions that keep the turf vigorous and attractive.

The Integrated Approach

Ultimately, the key to summer success lies in integration. By synergizing these strategies, you create a comprehensive maintenance program that adapts to the volatile summer environment. Detailed checklists, technological innovations, and continuous monitoring form the crux of an effective summer management plan. Each intervention, whether it's adjusting the irrigation schedule or rotating fungicides, contributes to a resilient turf that can withstand environmental stresses and continue to offer a superior playing surface.

Final Summary Checklist for Summer Maintenance & Heat Management

Turf Hydration Management:

Calibrate soil moisture sensors and adjust zonal irrigation.

Schedule early morning, deep watering sessions.

Use wetting agents and organic amendments to boost water retention.

Monitor ET data and adjust water budgets accordingly.

Pest and Disease Control:

Establish a regular scouting and monitoring protocol.

Implement cultural practices to reduce pest and disease risk.

Introduce beneficial insects and rotate chemical controls as needed.

Maintain detailed records on pest activity and treatment responses.

Heat Stress Prevention:

Increase mowing heights and maintain sharp mower blades.

Water early in the day with smart controllers to optimize performance.

Apply wetting agents and consider protective shading if necessary.

Regularly measure soil and turf temperatures for early warning signs.

Final Thoughts

Summer maintenance and heat management are not isolated tasks they are interwoven aspects of an integrated turf management system that responds dynamically to environmental challenges. Through precision irrigation, vigilant pest and disease management, and proactive measures against heat stress, you secure the course's long term performance and aesthetic appeal. A well executed summer strategy not only mitigates the negative effects of extreme weather but also sets the stage for continuous improvement throughout the season.

By adopting an integrated approach rooted in data, technology, and proven cultural practices, superintendents can transform the hottest, driest months into a period of sustained turf vigor and exceptional playability. The combined strategies detailed in this chapter empower you to face summer head on, ensuring that even during peak stress periods, your turf remains resilient, inviting, and pristine.

Chapter 6 Fall Transition & Preparing for Dormancy

Fall is a critical period for a golf course. Unlike the bustling schedules of spring and summer maintenance, the fall season is both a time for revitalization and an opportunity to prepare for the quiet months of winter. The work carried out during this period ranging from late season fertilization to meticulous debris removal has long term implications on turf health, playability, and overall course aesthetics. Additionally, thoughtful staffing adjustments help to manage costs and maintain operational efficiency during the winter months.

In this chapter, we focus on three major areas:

1. Late Season Fertilization for Root Development

2. Leaf Debris Removal and Turf Winterization

3. Planning Winter Staffing Adjustments

By addressing these components in detail, you will acquire the strategies and best practices necessary to transition from the high intensity summer operations to a period of dormancy with well prepared turf and streamlined operational logistics.

Late Season Fertilization for Root Development

The Role of Fertilization in Fall Transition

Fall fertilization is much more than a "last hurrah" before the dormancy period it is a strategic treatment designed to enhance root development, strengthen turf resilience, and prepare the soil for the stresses of winter. Unlike the fertilization programs of spring and summer, which focus on green up and rapid growth, fall treatments aim to promote deeper root growth that will allow the turf to better withstand the winter months and bounce back in the spring.

A well planned late season fertilization program typically emphasizes the following:

Enhanced Root Growth: Encouraging the development of a robust, deep root system ensures that the turf has the necessary reserves of carbohydrates and nutrients to survive dormancy.

Slow Release Nutrients: The use of slow release fertilizers provides a steady supply of essential nutrients over an extended period, translating into a gradual boost in turf health.

Balanced Nutrition: As the turf's above ground growth slows, directing nutrients toward root development is critical. A balanced ratio of nitrogen, phosphorus, and potassium (plus essential micronutrients) is key.

Best Practices in Late Season Fertilization

A. Timing and Application

Optimal Timing: Late fall (after the active growing season but before severe temperature drops) is ideal for fertilization. The goal is to apply nutrients while the soil is still warm enough to absorb them but before the turf goes completely into dormancy.

Sequential Applications: In some cases, split applications might be advantageous. An initial dose can jump start root growth, with a follow up to sustain nutrient release through the early winter months.

Application Methods: Uniform distribution is critical. Use calibrated spreaders and consider liquid applications in areas where the soil needs an immediate nutrient boost.

B. Nutrient Composition

Slow Release Nitrogen: Avoid high rates of readily available nitrogen; instead, focus on slow release products that minimize the risk of "luxury consumption" and excessive top growth.

Phosphorus and Potassium Boost: Both elements are crucial for root development. Phosphorus, in particular, fosters root establishment, while potassium plays a role in stress tolerance and overall plant metabolism.

Micronutrient Inclusion: Don't overlook micronutrients such as magnesium, iron, and zinc.

Even in minute amounts, these are essential for enzyme functions and overall turf vitality.

C. Integration with Soil Amendment Practices

Organic Matter: In areas where the soil lacks texture or organic content, incorporating compost or well rotted manure can improve the soil structure and water holding capacity.

PH Adjustment: Ensure that the soil's pH range is optimal for nutrient uptake (generally between 6.0 and 7.0 for most turf grasses). Liming or sulfur applications might be needed depending on the results of a soil test.

Real World Example

A superintendent in a northern climate noted that by implementing a dual phase fertilization program in the fall using a slow release nitrogen product combined with a phosphorus–potassium booster their greens exhibited a 25% increase in root biomass compared to previous seasons. The improved root system not only enhanced winter survival rates but also led to a quicker green up period in the succeeding spring.

Diagram Concept

Imagine a layered diagram that shows the root zone development over the fall season. The diagram can display:

-

The initial root system (depicted as shallow, sparse), then, after application of slow release fertilizer, a deeper, denser root structure.

Arrows indicating nutrient flow from the soil to the roots.

Boxes highlighting key ingredients (N P K ratios, micronutrients) and their effects.

Checklist Late Season Fertilization Essentials

Conduct a soil test in late summer to adjust nutrient plans based on current conditions.

Schedule fertilization for late fall when soil temperatures still allow nutrient uptake.

Use slow release fertilizers to promote sustained nutrient availability.

Apply balanced fertilizers with an emphasis on root promoting nutrients (phosphorus and potassium).

Incorporate soil amendments (organic matter, lime, or sulfur) as needed based on pH and texture.

Calibrate spreaders and verify even coverage across the entire course.

Document application rates and timings for future reference and adjustments.

Leaf Debris Removal and Turf Winterization

The Impacts of Fall Leaf Debris on Turf Health

As autumn progresses, trees and shrubs shed leaves that can quickly accumulate on the turf. While the fall colors provide picturesque scenes, leaf debris can pose significant challenges for turf maintenance:

Microscopic Fermentations: When left in large quantities, leaf layers trap moisture and promote fungal growth, including snow mold and other winter diseases.

Physical Barriers: A thick accumulation of leaves can smother the turf, restrict air circulation, and impede the penetration of winter treatments.

Delayed Dormancy: Excessive debris can alter the microclimate at the soil surface, potentially delaying dormancy and predisposing the turf to stress during sudden drops in temperature.

Strategies for Effective Leaf Debris Removal

A. Regular Collection

Mechanized Blowers and Vacuums: Invest in or rent equipment that can quickly remove fallen leaves from high traffic areas such as fairways, tee boxes, and around greens.

Manual Raking: In delicate areas especially on greens where mechanical devices might damage the turf manual removal using rakes or specialized leaf clearing tools is recommended.

B. Timing of Removal

Pre Winter Cleanup: Aim to complete major debris removal before the onset of winter weather. Late autumn, after the bulk of leaves have fallen, is the ideal time to clear the course.

Frequent Monitoring: Regularly check and remove fresh debris throughout the late season. In regions with rapid leaf drop, daily quick clean routines may be necessary for high traffic areas.

Turf Winterization Techniques

A. Preparation for Dormancy

Reduced Mowing and Lower Traffic: Gradually reduce mowing frequency as the turf approaches dormancy. Lowering traffic reduces stress and helps the turf settle into a stable state for winter.

Irrigation Adjustments: The need for water decreases as temperatures drop. Begin tapering irrigation schedules to avoid waterlogging and subsequent freeze thaw damage.

Application of Winterizers: Winterizer fertilizers typically have a higher potassium content, which helps strengthen cell walls and improves the turf's ability to withstand cold stress.

B. Preventative Disease Control

Fungicide Applications: In climates where snow mold or other winter fungal diseases are problematic, a prophylactic fungicide application in the fall can be an effective preventive measure.

Post Debris Treatment: After removing leaf debris, consider light topdressing or a gentle de thatching process to ensure the turf has a clear, unobstructed surface.

Real World Example

A superintendent in a northern golf course region implemented an aggressive leaf removal program using a combination of blower vacuums for fairways and manual raking for greens. This systematic approach, paired with a robust winterizer fertilization program, reduced snow mold incidence by 40% compared to previous seasons. The improved air circulation and nutrient balance translated into a uniform, dense turf that withstood harsh winter conditions with minimal patchiness.

Diagram Concept

Visualize a "before and after" schematic that displays a turf area with heavy leaf cover versus one that has been properly cleared. The diagram can highlight:

The layers of leaf debris with potential microclimate effects (moisture retention and limited airflow).

The use of equipment (blowers, vacuums, rakes) to remove debris.

The subsequent application of topdressing or winterizer fertilizer to prepare the turf for dormancy.

Checklist Leaf Debris Removal & Winterization Essentials

Monitor leaf fall regularly and remove debris promptly.

Deploy mechanical or manual removal methods appropriate for different areas of the course.

Complete a major debris removal cleanup in late autumn, ideally before the first hard frosts.

Adjust mowing operations to reduce turf disturbance and avoid creating thatch or compaction.

Begin tapering irrigation schedules to prevent waterlogging as temperatures drop.

Apply winterizer fertilizers with a high potassium component.

Consider prophylactic fungicide treatments to protect against winter fungal diseases.

Document removal routines and adjust frequencies based on weather and leaf fall patterns.

Planning Winter Staffing Adjustments

The Importance of Strategic Winter Staffing

While fall is a season of harvesting the results of a busy year, it is also the time to reassess workforce needs and prepare for the slower pace of winter operations. Winter staffing adjustments are essential for both cost management and ensuring that the core functions of the course are maintained at a sustainable level during the dormancy period. Strategic planning in this area not only involves trimming down the workforce but also reallocating and retraining personnel so that operations can resume seamlessly in the spring.

Key Considerations for Winter Staffing

A. Reduced Operational Demands in Winter

Lower Course Activity: Golf courses experience a significant drop in play during winter, reducing the need for full scale maintenance operations.

Prioritization of Critical Tasks: Focus on essential functions such as equipment maintenance, facility repairs, and winter specific treatments rather than full scale turf management.

Cost Efficiency: Reducing staffing levels and scaling back operations can help control overhead costs during months when revenue generation is limited.

B. Reallocation and Cross Training of Staff

Multi Skilled Teams: Cross training staff on multiple core areas such as equipment servicing, facility upkeep, and administrative functions ensures that a lean team can cover essential tasks without the need for a large workforce.

Rotation Schedules and Shifts: Develop staff rotation schedules that allow for continuous coverage of critical maintenance tasks while giving employees adequate rest and reducing burnout.

C. Preparation for Spring Re Activation

Training and Development: Use the winter months as an opportunity to invest in staff training. Organize workshops, certification programs, and refresher courses on the latest turf management technologies and safety protocols.

Staff Meetings and Reviews: Hold year end review sessions to assess performance, discuss improvements, and plan for the upcoming season. Incorporate staff feedback to refine operational strategies and identify potential new hires or adjustments needed for the spring season.

Developing a Strategic Winter Staffing Plan

Analyzing Operational Needs

Mapping Tasks: List all functions that must continue during winter (e.g., snow removal if applicable, equipment servicing, routine facility checks) and determine the number of staff required for each.

Critical vs. Non Critical Tasks: Identify areas where work can be postponed until spring versus those that truly need winter attention. This helps in deciding where staffing can be temporarily reduced.

Communication and Transition Strategies

Transparent Communication: Inform staff well in advance about offseason staffing adjustments, potential layoffs, or temporary reassignments.

Providing clarity helps maintain morale and sets expectations.

Severance and Rehiring Plans: If reductions are necessary, develop clear policies for severance, benefits, and the rehiring process once the busy season resumes. Consider offering training opportunities during the winter to support earlier returns.

Technological and Administrative Aids

Scheduling Software: Utilize digital scheduling tools to streamline winter assignments and ensure coverage of critical functions.

Performance Monitoring: During the reduced winter period, maintain performance logs and feedback systems to identify areas where the reduced team might need additional support or adjustment.

Real World Example

A superintendent facing an exceptionally mild winter implemented a dynamic staffing plan that reduced full time staff by 40% during the dormancy period. By cross training remaining employees on both equipment maintenance and administrative duties and incorporating remote monitoring tools for field checks, the course maintained essential functions. Moreover, yearend training workshops provided staff with new skills that led to a smoother, more efficient spring re activation, with higher employee satisfaction scores in subsequent seasons.

Diagram Concept

Envision a flowchart that outlines the winter staffing adjustment process:

Start with an assessment of winter operational needs,

Branch into critical vs. non critical tasks,

Define staffing levels for each category,

Then show communication and training initiatives,

Finally, transition planning with timelines for reactivation in the spring.

Checklist Winter Staffing Adjustments

Conduct an operational audit to map necessary winter tasks.

Identify critical functions that require full coverage versus those that can be deferred.

Develop a clear, documented staffing plan that reduces workforce as needed.

Cross train remaining staff on essential multi functional roles.

Communicate staffing changes and expectations to all team members well in advance.

Plan and schedule winter training workshops and review sessions.

Use scheduling and performance software to monitor workflow and identify gaps.

Establish protocols for rehiring and transition back to full operations in the spring.

Conclusion: Integrating Fall Transition Strategies for Long Term Success

Fall transition and preparation for dormancy are pivotal for setting the stage for a successful winter and a vigorous restart in the spring. The steps outlined in this chapter, from late season fertilization to meticulous leaf removal and strategic winter staffing adjustments, are not isolated tasks; they are interconnected practices that ensure the longevity and resilience of both the turf and the overall operational framework of your golf course.

Late season fertilization directs crucial nutrients to the roots, thereby equipping the turf with the means to endure the winter's challenges. At the same time, effective debris removal and targeted winterization practices create a healthier, disease free environment that enhances the turf's natural defenses. Finally, planned and thoughtful staffing adjustments provide not only immediate cost savings but also a prepared and well trained team ready to ramp up operations when warmer weather returns.

Together, these integrated strategies form the foundation for an avenue of continuous improvement. Fall is the season to reflect on past successes, identify areas for strategic refinement, and implement practices that will yield a resilient, thriving turf in the spring. By establishing comprehensive checklists, leveraging technology, and embracing both cultural and operational innovations, superintendents can ensure that their courses transition smoothly through dormancy, emerge stronger, and retain their competitive edge year after year.

Final Summary Checklist for Fall Transition & Winter Preparation

Late Season Fertilization for Root Development:

Conduct soil tests in late summer for nutrient recalibration.

Schedule fertilization before the onset of deep winter.

Use slow release fertilizers formulated for root enhancement.

Integrate organic amendments and adjust pH as needed.

Record and review application data for future adjustments.

Leaf Debris Removal and Turf Winterization:

Establish a regular leaf removal routine using mechanical and manual methods.

Clear major infill areas before frost sets in.

Adjust mowing frequencies and heights to reduce stress.

Taper irrigation schedules and apply winterizer treatments.

Consider prophylactic fungicide applications to prevent snow mold.

Document clearance schedules and measure turf improvement metrics.

Planning Winter Staffing Adjustments:

Audit essential functions and map out winter tasks.

Identify critical operations that require dedicated personnel.

Develop cross training programs to streamline responsible roles.

Communicate changes, set expectations, and prepare reactivation protocols.

Use digital tools for scheduling, tracking, and performance analysis.

- Plan year end training and review sessions for a seamless transition.

Final Thoughts

The fall transition period is a time of strategic consolidation and proactive preparation. While it may appear to be a phase of reduced activity, the careful actions taken during this time have pronounced long term benefits. By focusing on targeted fertilization that nourishes the roots, diligent debris removal to eliminate potential problems, and well planned staffing adjustments that set the stage for operational efficiency, you secure the health of your turf and the integrity of your course.

Embracing these comprehensive strategies ensures that your golf course enters the winter months not as a dormant entity, but as a resilient, well prepared system ready for renewal the following spring. In the cycles of nature, the foundation laid during fall is the bedrock upon which future growth is built. When every leaf is cleared, every nutrient applied, and every staffing decision made with foresight, you create the conditions for a seamless transition and ultimately, a superior playing experience for seasons to come.

Chapter 7: Winter Strategy & Long Term Planning

As winter descends upon a golf course, the pace of operations shifts dramatically from peak maintenance to preservation and strategic planning. Although the course may appear dormant on the surface, this season offers a crucial window for in depth equipment maintenance, disease prevention, and thoughtful reflection on the previous year's operational challenges. A well executed winter strategy not only protects the physical assets during harsher conditions but also provides a strong foundation upon which to build the coming season's successes.

In this chapter, we address three highly interrelated components of winter management:

1. Equipment Servicing and Repair Schedules: Ensuring that every piece of machinery and critical infrastructure is thoroughly inspected, maintained, and repaired during the off season.

2. Snow Mold Prevention Treatments: Implementing strategies to mitigate the risk of snow mold a common disease during winter that can weaken turf and jeopardize spring recovery.

3. Reviewing Past Season Challenges: Conducting a comprehensive debrief of the previous season's successes and shortcomings to inform long term planning and operational improvements.

Each section provides detailed strategies, step by step processes, diagram ideas, and checklists to guide superintendents through the complex landscape of winter operations and long term strategic planning.

Equipment Servicing and Repair Schedules

The Critical Role of Winter Equipment Maintenance

Winter is the ideal time to perform in depth servicing and repair on all essential equipment. With the course operating on a reduced schedule, superintendents have the opportunity to schedule repairs, calibrations, and preventative maintenance that would be too disruptive during the busy growing season. This proactive approach ensures that machinery functions reliably when the course reactivates in the spring, reducing the likelihood of breakdowns and extending the lifespan of expensive assets.

Developing a Comprehensive Servicing Schedule

A. Establishing a Preventative Maintenance Calendar

Inventory Review: Begin by compiling a comprehensive inventory of all equipment from mowers, aerators, and tractor fleets to irrigation and facility maintenance vehicles. Include all critical components such as engine parts, hydraulic systems, and electrical controls.

Manufacturer Guidelines: Consult the maintenance and warranty guidelines provided by manufacturers. These documents offer recommended servicing intervals, parts replacement schedules, and calibration procedures that are critical for maintaining peak performance.

Seasonal Priorities: Group equipment into categories based on usage intensity. For instance, machinery used daily on greens may require more frequent servicing compared to equipment used occasionally on maintenance tasks elsewhere on the course.

B. Scheduling Specific Repairs and Servicing Tasks

Pre Season vs. Off Season: Identify tasks that must be completed before the first play of the new season, such as blade sharpening and calibration of core mowers. Schedule more extensive repairs and overhauls for the off season when time and weather permit.

Detailed Task Lists: For each piece of equipment, create a detailed list outlining:

Daily or weekly checks (visual inspections, fluid levels, minor adjustments)

Monthly deep cleaning and diagnostics

Seasonal tasks (engine overhauls, replacing worn parts, calibrations)

Integration with Digital Tools: Consider using maintenance management software or digital scheduling tools that send alerts when maintenance deadlines are approaching. These systems help keep records consistently updated and can identify recurring issues.

C. Record Keeping and Data Analysis

Maintenance Logs: Document every service activity with dates, actions taken, parts replaced, and any anomalies observed. Over time, these records become invaluable for identifying patterns, budgeting for future repairs, and negotiating service contracts.

Performance Audits: Regularly review performance data from equipment sensors (if available) and feedback from operators. This data driven approach helps refine maintenance schedules and predict which components are most likely to require attention in the coming year.

Training and Safety Reviews: Use the off season as an opportunity to review safety protocols and provide refresher training for operators and maintenance staff. Integrating safety checks into the maintenance schedule not only protects personnel but also contributes to the longevity of the machines.

Real World Example

A superintendent at a multi course facility implemented a structured winter maintenance program using both physical logbooks and digital maintenance management software. Over three winter seasons, this program reduced unscheduled downtime by nearly 35%, extended the service life of several key pieces of equipment, and allowed the facilities team to negotiate better service contracts by showing historical data of proactive maintenance interventions.

Diagram Idea

Consider a Gantt chart or a timeline diagram that outlines the entire off season maintenance schedule. This visual could detail daily, weekly, and seasonal tasks for different equipment categories, clearly marking deadlines and priority items. Such a diagram serves as a practical guide that can be updated yearly and shared with the maintenance team.

Checklist Equipment Servicing and Repair Essentials

Compile a complete inventory of all essential equipment and infrastructure.

Consult manufacturer maintenance guidelines and create a detailed servicing calendar.

Schedule regular visual inspections, cleaning, and fluid level checks.

Plan and document monthly, seasonal, and pre season maintenance tasks.

Utilize digital scheduling tools to manage maintenance deadlines.

Maintain detailed logs of all repair and maintenance work.

Perform operator and safety training as part of the off season routine.

Review historical maintenance data to inform future schedules.

Snow Mold Prevention Treatments

Understanding the Threat of Snow Mold

Snow mold is a common fungal disease that affects turf during the winter months, particularly in climates where snow cover and prolonged periods of cold, moist conditions prevail. There are two main types of snow mold: gray snow mold and pink snow mold. Both can damage the turf by causing discoloration and weakening the plant tissues even if the turf appears to recover in the spring, the underlying root systems may be compromised, leading to uneven playability and delayed green up.

Identifying Snow Mold and Its Symptoms

A. Visual and Physical Signs

Gray Snow Mold: Typically forms under prolonged snow cover in wet conditions. The affected areas may appear irregularly shaped, with a grayish hue after the snow melts.

Pink Snow Mold: Often begins with pinkish discoloration that rapidly turns to a white or grayish mass. This type is generally more aggressive and can spread quickly in humid conditions.

Subtle Indicators: Frequently, early snow mold is not immediately obvious. Look for areas where turf seems less dense, or where leaf blades are discolored even before complete snow melt.

Preventative Strategies for Snow Mold

A. Cultural Management Practices

Turf Hygiene:

Leaf and Debris Removal: Ensure that the turf is cleared of excessive leaf debris before the onset of heavy snow. Debris can trap moisture and create the ideal microenvironment for fungal growth.

Aeration and Thatch Control: Prior fall aeration activities contribute to better air circulation under the snow cover, reducing the disease favorable conditions.

Balanced Fertilization:

Low Nitrogen Winterizer: A winterizer fertilizer with a higher proportion of potassium and controlled nitrogen releases can deter the overgrowth that predisposes turf to infection.

Avoid Over Fertilization in Late Fall: Excess nutrients can encourage lush growth, which is more prone to infection and less tolerant of cold.

Irrigation Management:

Reduce Precipitation Dependence: Minimize irrigation towards the end of the fall season when natural precipitation and snow are imminent. Over watering can lead to excessive moisture in turf tissues.

B. Chemical Controls

Fungicide Applications:

Timely and Targeted Treatments: If historical data or early reports indicate a high risk for snow mold, plan a prophylactic fungicide treatment in the late fall. This application is generally made after significant leaf and debris management.

Fungicide Rotation: Use a rotation of fungicides to prevent the development of pathogen resistance. Follow label guidelines carefully, particularly regarding timing and dosage.

Environmental Considerations:

Weather Dependent Timing: Fungicide applications must consider upcoming weather patterns. Avoid treatments immediately before heavy snowfall unless absolutely necessary, as coverage may be reduced by snow melt.

Monitoring and Early Detection

Regular Inspections: Once the snow begins to melt, inspect key areas of the course for signs of snow mold. Early detection allows for timely remediation before the disease spreads.

Record Weather Impact: Maintain records of winter weather conditions, snow cover duration, and any observed disease symptoms. This data will support future decision making and preventative measures.

Collaboration with Experts: Engage with local extension services or turf grass pathologists to assess the risk and adapt treatment protocols as needed.

Real World Example

A golf course in a region prone to heavy winter snowfall established a preemptive protocol for snow mold prevention. By implementing a rigorous debris removal schedule in late autumn, followed by a targeted fungicide application based on weather forecasts, the course reduced the incidence of snow mold by 50% over three successive winters. This not only improved spring recovery rates but also reduced subsequent fungicide costs by minimizing the need for curative treatments.

Diagram Idea

Develop a schematic flowchart that outlines the snow mold prevention cycle: starting with cultural practices (debris removal, aeration, and balanced fertilization), moving into the criteria for fungicide application, and concluding with post snow melt inspections. The diagram should visually represent the interplay between preventative measures and early treatment interventions.

Checklist Snow Mold Prevention Essentials

Conduct thorough leaf and debris removal in late autumn.

Schedule fall aeration and thatch control to improve under snow airflow.

Adjust winterizer fertilizer programs to avoid excessive nitrogen.

Monitor local weather forecasts and adjust irrigation practices accordingly.

Plan and execute a timely fungicide application based on risk assessment.

Rotate fungicide products to prevent resistance development.

Inspect turf immediately following snow melt for early signs of infection.

Document weather data, treatment details, and disease incidence for future reference.

Reviewing Past Season Challenges

The Importance of Reflective Practice in Long Term Planning

Reviewing the past season's challenges is a critical component of long term planning. By taking a systematic approach to analyzing successes and setbacks, superintendents can fine tune maintenance practices, optimize budgets, and implement innovative strategies that address recurring issues. This reflective practice is not merely a retrospective exercise; it lays the groundwork for future operational improvements and increased resilience.

Key Areas for Post Season Review

A. Equipment and Infrastructure Performance

Maintenance Records: Analyze equipment servicing logs, downtime, and repair costs. Identify recurring issues and note any deviations from the planned maintenance schedule.

Operational Efficiency: Evaluate the performance of key machinery in relation to the preventive maintenance strategies implemented during the season. Were there unexpected breakdowns, and if so, what were their root causes?

Capital Expenditures: Review major expenditures on equipment and infrastructure upgrades. Determine if investments yielded the expected improvements and if additional upgrades might be required.

B. Turf Health and Cultural Practices

Disease Pressure Analysis: Review records of pest and disease outbreaks over the season including timing, severity, and geographical spread and correlate these with factors such as irrigation, fertilization, and mowing practices.

Water Management: Evaluate water usage based on sensor data and irrigation logs. Assess whether adjustments in watering schedules adequately met the turf's needs and where inefficiencies remain.

Growth and Recovery: Compare data on turf density, color, and recovery rates post interventions. Identify if earlier season treatments (e.g., winterizer fertilizations or pre emergent applications) had the desired long term effects.

C. Workforce and Operational Adjustments

Staffing Efficiency: Review winter and summer staffing decisions to determine whether the adjustments effectively balanced cost control with operational readiness.

Training and Development: Evaluate the impact of any training sessions or cross training efforts on performance and overall team efficiency.

Feedback Mechanisms: Gather qualitative feedback from staff regarding operational hurdles, safety concerns, and suggestions for improvement. Incorporate these insights into the strategic planning process.

Tools and Techniques for Reviewing Season Data

Surveys and Interviews: Conduct post season surveys with key personnel and maintenance staff to collect insights that quantitative data might not fully reveal.

Data Analytics: Leverage digital tools to aggregate maintenance logs, water usage reports, and equipment performance metrics into clear visual reports. Use these analytics to spot trends and anomalies.

SWOT Analysis: Perform a Strengths, Weaknesses, Opportunities, and Threats (SWOT) analysis regarding your course's performance over the season. This framework helps contextualize operational outcomes in a broader strategic setting.

Benchmarking: Compare seasonal data with historical records or industry benchmarks. Identifying deviations from expected parameters can prompt a review of both operational protocols and external market conditions.

Real World Example

A superintendent implemented an annual review process that combined digital performance tracking with structured staff interviews. The process identified a pattern of increased equipment downtime during peak summer months, correlating strongly with an unplanned spike in irrigation system failures. These findings led to a targeted update of the irrigation maintenance schedule and further training for operators. As a result, subsequent seasons showed both a reduction in unplanned downtime and improved overall turf performance.

Diagram Idea

A flow diagram outlining the season review process could start with data collection (maintenance logs, water usage, staff surveys), followed by data analysis (trend identification, SWOT analysis), and concluding with actionable items for next season adjustments. This visual tool can be part of an annual review presentation for stakeholders.

Checklist Reviewing Past Season Challenges

Collect and consolidate all maintenance and equipment logs.

Analyze water usage data and correlate with turf performance.

Review pest and disease outbreak records alongside cultural practice records.

Conduct surveys and interviews with staff to gather qualitative insights.

Perform a SWOT analysis of the season's operations.

Identify and prioritize actionable improvements based on data insights.

Document key findings and incorporate them into next season's planning schedule.

Conclusion: A Foundation for Future Excellence

Winter Strategy & Long Term Planning is not simply about surviving the off season but about laying a robust foundation for continuous improvement. By prioritizing thorough equipment servicing, combating winter diseases like snow mold, and conducting an honest review of the previous season's challenges, you create a systematic framework that drives both immediate operational readiness and long term success.

A course that demonstrates excellence in winter maintenance is one that can emerge more resilient and perform at a superior level in the spring. With rigorously maintained machinery, a proactive approach to disease prevention, and an analytical review process that transforms past challenges into strategies for future growth, superintendents are poised to excel in every season.

Final Summary Checklist for Winter Strategy & Long Term Planning

Equipment Servicing and Repair:

Complete a comprehensive equipment inventory and schedule servicing.

Follow manufacturer guidelines and integrate digital maintenance tools.

Document all repairs and calibrations for future trend analysis.

Provide operator training and conduct regular safety reviews.

Snow Mold Prevention Treatments:

Remove debris and manage turf conditions in late fall.

Adjust fertilization and irrigation schedules to deter fungal growth.

Plan timely prophylactic fungicide applications when high risk is forecast.

Monitor spring emergence for early signs of snow mold and document findings.

Reviewing Past Season Challenges:

Aggregate maintenance, water, and pest/disease performance data.

Conduct SWOT analyses and staff surveys for qualitative insights.

Benchmark current season performance against historical data.

Develop actionable improvements for updating future operational plans.

Final Thoughts

A successful winter strategy and effective long term planning are indispensable pillars of a well managed golf course. By using the winter months to rigorously maintain equipment, preemptively address snow mold issues, and critically review the lessons learned over the past season, you not only safeguard your assets but also chart a clear course for future growth. This integrated approach ensures that when spring arrives, every component from machinery to turf health and workforce capability is primed for a robust comeback.

Remember, the winter season is as strategic as it is dormant. Embrace this opportunity to refine your operational practices, document every lesson, and invest in both technological and human capital. By doing so, you transform inevitable challenges into stepping stones toward sustained excellence and competitive advantage in the seasons to come.

Chapter 8: Common Turf grass Problems & Solutions

Turf grass management is as much about solving problems as it is about preventative care. Despite best efforts, issues such as fungal diseases, weed outbreaks, invasive pest pressures, and drainage challenges can arise. This chapter provides a comprehensive guide to diagnosing and tackling these common problems, ensuring your golf course not only recovers swiftly from disturbances but maintains long-term playability and visual appeal.

The chapter is organized into three primary sections:

1. Diagnosing Fungal Diseases and Weed Outbreaks

2. Managing Invasive Pests

3. Repairing Bunker Drainage Issues

Each section is designed to provide detailed operational techniques, enabling you to pinpoint issues early and respond with targeted solutions.

Diagnosing Fungal Diseases and Weed Outbreaks

Understanding the Problem

Fungal diseases and weed infestations are two of the most persistent challenges on turf grass, with causes that often overlap. Warm, humid conditions in summer and extended damp periods in spring or fall create a favorable environment for pathogens and undesirable plants alike. Without prompt diagnosis and intervention, these issues can reduce turf density, compromise playability, and lead to increased maintenance costs.

Fungal Diseases: Identification and Common Causes

A. Common Fungal Diseases on Turf grass

Dollar Spot: Characterized by small, silver dollar-sized lesions on the turf blades, Dollar Spot is one of the most frequently encountered diseases on golf courses. Early signs include tiny yellow spots that can merge into larger brown patches if untreated.

Brown Patch: Often affecting cool season grasses, Brown Patch appears as circular, brown patches with dark borders. It typically erupts on high use areas, especially when nights are cool and days are warm.

Powdery Mildew: Appearing as a white, powdery substance on the leaf surfaces, Powdery Mildew usually occurs during periods of low humidity or when excessive nitrogen has spurred lush but vulnerable growth.

B. Diagnostic Techniques for Fungal Diseases

Visual Inspection: Walk the course systematically looking for changes in turf color, texture, and density. Document areas that exhibit unusual discoloration, water soaked appearances, or signs of blighting.

Sample Collection and Laboratory Analysis: For persistent or ambiguous issues, collect turf samples from affected and non-affected zones. Laboratory tests can confirm pathogen identity and guide treatment recommendations.

Environmental Assessment: Record local weather data, soil moisture levels, and recent management practices. Many fungal diseases are exacerbated by over irrigation, poor drainage, or inadequate airflow.

Weed Outbreaks: Diagnosis and Causes

A. Identifying Weed Outbreaks

Symptoms: Weeds often present as sudden mounds or patches of plants that contrast with the uniform turf grass. Look for variations in leaf shape, color, and growth habit.

Causal Factors: Weak turf, over fertilization, or poor irrigation practices can give weeds a competitive edge. Additionally, disturbed soil from aeration or construction can create ideal conditions for weed seed germination.

B. Distinguishing Between Fungal Diseases and Weeds

Leaf Texture and Growth Form: Fungal symptoms typically appear as discolorations or lesions on existing grass blades, whereas weeds will have a distinct growth form and structure.

Pattern of Occurrence: Disease outbreaks tend to follow patterns associated with moisture and temperature stress, whereas weed emergence may be more sporadic, often following disturbances in the turf.

Integrated Solutions for Fungal Diseases and Weed Outbreaks

A. Cultural Practices

Optimized Irrigation: Use precise irrigation scheduling and zoning to prevent prolonged leaf wetness that promotes fungal growth and to discourage weed seed germination.

Proper Mowing and Thatch Management: Maintain appropriate mowing heights to reduce canopy density and improve airflow. Regular dethatching prevents the accumulation of organic material that serves as a breeding ground for pathogens and weeds.

Balanced Fertilization: Avoid excessive nitrogen that can lead to overly lush, tender growth. Instead, apply balanced, slow release formulations that foster steady, robust turf without inviting disease or weed infestations.

B. Chemical Controls

Fungicides: Apply fungicides preventatively when environmental conditions predispose turf to fungal infections. Follow label recommendations carefully and rotate products to prevent resistance.

Herbicides for Weed Control: Use selective herbicides to target weed species without damaging the turf. Spot treatments are often more effective than broad applications, especially in areas of localized infestation.

Real World Example

At one high end urban golf course, a superintendent began a dual pronged approach after noticing patchy, discolored areas on the greens. Through detailed mapping, laboratory labs were used to confirm early Dollar Spot infection. Concurrently, an area of emerging weed infestation was noted in a corner of the course where turf was thin from compaction. By adjusting the irrigation schedule, applying a targeted fungicide, and spot treating the weed patches, the course maintained uniform turf quality and reduced the need for extensive later season interventions.

Diagram Idea

Imagine a diagnostic flowchart that begins with "Visual Inspection" then branches into "Fungal Disease Symptoms" and "Weed Characteristics." Each branch should include key indicators (e.g., lesion size, turf discoloration, leaf structure) and follow with recommended diagnostic steps such as sample collection, environmental data logging, and specific treatment interventions.

Comprehensive Checklist Diagnosing Fungal Diseases and Weed Outbreaks

Conduct systematic visual inspections across all turf zones.

Record signs of discoloration, lesions, and abnormal growth patterns.

Collect samples from affected and unaffected areas for comparative analysis.

Monitor local weather, soil moisture, and recent management actions.

Identify differences between fungal symptoms and weed growth forms.

Adjust cultural practices (irrigation, mowing, and fertilization) based on findings.

Implement targeted fungicide or herbicide treatments as needed.

Document and map areas of recurring problems for future reference.

Managing Invasive Pests

Understanding the Impact of Invasive Pests

Invasive pests represent a continuous threat to turf health by directly feeding on plants and indirectly facilitating the spread of diseases. Invasive species often proliferate when native predators are absent, or when management practices inadvertently favor rapid growth of pest populations.

A. Common Invasive Pests on Golf Courses

Chinch Bugs: These pests feed on the sap of turf grass and can cause significant damage by creating localized dead patches on greens and fairways.

Sod Webworms: Feeding on both the blades and roots, sod webworms can lead to widespread thinning of the turf in severe infestations.

Invasive Grubs and Larvae: Soil dwelling grubs and larvae feed on the roots of turf grass, undermining the structural integrity of the plant and leading to sudden turf loss in affected areas.

Diagnostic and Monitoring Strategies for Invasive Pests

A. Regular Scouting and Field Inspections

Visual Surveys: Inspect turf regularly especially during peak pest activity times for signs of damage such as browning, wilting, or unusual patterns of turf loss.

Soil Sampling: Conduct soil sampling to detect the presence of grubs or other larvae. Look for feed tunnels or areas of discolored, dying turf.

B. Use of Traps and Pheromones

Sticky Traps and Pheromone Lures: These tools can provide early warnings about pest population increases. Position them in areas prone to pest pressure and monitor them frequently.

Data Logging: Record pest counts and associated damage levels. Over time, this data will help you set thresholds for initiating control measures.

Integrated Pest Management (IPM) Solutions

A. Cultural Controls

Soil and Turf Health: Healthy, well maintained turf is less susceptible to pest infestations. Optimize nutrient management, proper irrigation, and aeration to support vigorous growth and natural resistance.

Resistance through Diversity: Implementing a mix of turf grass species can sometimes reduce the likelihood of widespread pest outbreaks, as certain species may be less palatable to invasive pests.

B. Biological Controls

Beneficial Insects: Introduce or encourage populations of natural predators (such as parasitic wasps) that feed on pest species.

Nematodes and Microbial Agents: Biological pest control products, which include beneficial nematodes, can help to control soil dwelling pests such as grubs without harming the turf.

C. Chemical Controls

Targeted Insecticides: Use insecticides judiciously and only when pest counts reach established economic thresholds. Rotate products regularly to reduce the risk of developing resistance.

Spot Treatments: In cases where only a small area is affected, localized chemical treatments are preferable to ensure that the overall ecosystem is minimally disrupted.

Real World Example

A golf course in a moderately warm climate experienced a sudden outbreak of chinch bugs in one of its fairways. The superintendent intensified his scouting regimen and installed sticky traps. When pest counts exceeded predetermined thresholds, a spot insecticide treatment was applied while complementary cultural practices (such as adjusting irrigation schedules and improving mowing heights) were also implemented. Over the subsequent weeks, pest populations were controlled without widespread chemical use, and the turf quickly recovered.

Diagram Idea

Develop a layered diagram that illustrates the integrated pest management cycle:

1. **Initial Scouting and Monitoring** (visual inspections, traps)

2. **Diagnosis** (identifying pests, quantifying damage)

3. **Cultural & Biological Controls** (maintaining healthy turf, introducing beneficials)

4. **Chemical Intervention (if necessary)**

5. **Follow Up Monitoring and Data Collection.**

Comprehensive Checklist Managing Invasive Pests

Establish routine visual surveys in high risk areas.

Install sticky traps and pheromone lures at strategic locations.

Conduct soil sampling to detect early infestations by grubs and larvae.

Record pest counts and correlate with turf damage observations.

Implement cultural practices that strengthen turf resilience.

Introduce beneficial insects or microbial agents as part of biological control.

Apply chemical treatments only when necessary and rotate insecticide classes.

Follow up with regular scouting to assess the effectiveness of interventions.

Repairing Bunker Drainage Issues

The Critical Role of Bunker Drainage

Bunkers are not only visually distinctive features on a golf course but also essential technical elements that affect player safety and the course's overall aesthetics. Proper drainage in bunkers ensures that water does not accumulate, which can lead to soft, unstable footing and an increased risk of erosion. Poor bunker drainage also creates potential breeding grounds for invasive weeds and diseases due to prolonged moisture accumulation.

Diagnosing Bunker Drainage Problems

A. Visual and Physical Inspections

Water Pooling: Look for signs of standing water or areas where water consistently collects after rainfall. Pay particular attention to low points in the bunker.

Soil Saturation and Compaction: Inadequate drainage can result in waterlogged, compacted soil that inhibits grass recovery and accelerates wear. Use soil probes to test the infiltration rate in problematic zones.

Surface Erosion and Cracking: Bunker surfaces may show signs of erosion or cracking in areas with persistent drainage issues. These are indicators that the subsoil or drainage infrastructure may require attention.

Repair and Preventative Strategies for Bunker Drainage

A. Structural Repairs and Regrading

Re Leveling the Bunker: Evaluate the slope and grading of the bunker. If water is pooling, re grading may be necessary to enhance natural water runoff.

Installing Additional Drainage: In severe cases, consider incorporating supplemental drainage solutions such as:

French Drains: Trenches filled with gravel and perforated pipes that direct excess water away from critical areas.

Subsurface Drainage Layers: Installing layers of specialized drainage material (e.g., geotextile fabrics coupled with aggregate layers) underneath the bunker can enhance water permeability.

Periodic Maintenance: Just like turf grass, bunker drainage systems require regular maintenance. Clearing blockages from drainage channels and resealing surfaces can prevent minor issues from becoming major concerns.

B. Cultural Controls to Complement Structural Repairs

Grass Selection and Turf Maintenance: Choose turf grass varieties that are particularly tolerant of moisture and that can quickly recover from wet conditions. Regular mowing and aeration around bunkers can also improve surface drainage.

Organic Matter Amendments: Incorporate organic matter into the bunker's soil to improve structure and permeability, ensuring that water flows through rather than sitting on the surface.

Real World Example

At one prestigious country club, persistent water pooling had created soft spots on bunkers, leading to excessive wear and player dissatisfaction. A comprehensive review of the bunker's grading revealed subtle depressions and blockages in the drainage system. The superintendent re graded the bunkers, installed French drains along the low points, and introduced a scheduled maintenance routine for periodic debris removal. The result was a dramatic improvement in bunker playability and a reduction in long term repair costs.

Diagram Idea

Create a cross sectional diagram of a bunker showing:

The surface grading and topsoil layer.

The drainage system below, including French drains or perforated pipe installations.

Arrows depicting the intended flow of water away from the bunker. This visual aid can help explain both the diagnosis and the remediation process.

Comprehensive Checklist Repairing Bunker Drainage Issues

Inspect bunkers for signs of water pooling, erosion, and soil compaction after rainfall.

Use soil probes to determine the infiltration rate in problem zones.

Evaluate the grading of each bunker; determine if re leveling is required.

Develop a plan for supplemental drainage installations (such as French drains or subsurface layers).

Schedule regular maintenance to clear blockages and inspect drainage channels.

Document repairs and monitor impacted areas for post repair performance.

Integrate cultural practices (aeration, organic amendments) that complement drainage improvements.

Conclusion: Integrated Solutions for Turf grass Challenges

Effective turf grass management requires a keen understanding of the common problems that can undermine your course's performance and appearance. By diagnosing and addressing fungal diseases, weed outbreaks, invasive pest infestations, and bunker drainage issues, you protect and enhance your turf's resilience throughout the year.

Key Takeaways:

Fungal Diseases and Weed Outbreaks: Early and accurate diagnosis supported by visual inspections, sample analysis, and environmental assessments is critical. Use an integrated management approach that combines cultural, biological, and chemical strategies to restore and maintain turf health.

Invasive Pests: Vigilant scouting and monitoring, coupled with an integrated pest management (IPM) plan, help keep pest populations under control. Balancing cultural improvements with targeted biological and chemical interventions minimizes long term turf damage.

Bunker Drainage: For specialized areas like bunkers, proper drainage is paramount. Structural repairs such as regrading and the installation of additional drainage systems complement cultural practices that improve soil structure and turf recovery.

By combining these solutions within a coordinated management program, you not only address immediate turf challenges but also lay the groundwork for long term success. Each strategy, checklist, and diagram presented in this chapter forms part of an integrated system that bridges diagnosis with effective remediation ensuring that your course remains a model of operational excellence.

Final Summary Checklist for Common Turf grass Problems & Solutions

Diagnosing Fungal Diseases and Weed Outbreaks:

Conduct systematic field inspections.

Record symptoms and compare affected versus healthy turf.

Collect samples for laboratory analysis if issues persist.

Adjust irrigation, mowing, and fertilization practices to reduce disease pressure.

Use targeted fungicide and herbicide treatments based on accurate diagnosis.

Managing Invasive Pests:

Set up regular scouting routines and use traps/pheromone lures.

Record and analyze pest population data.

Implement cultural and biological controls as first line measures.

Apply insecticides only when economic thresholds are exceeded.

Rotate chemical classes to prolong effectiveness.

Repairing Bunker Drainage Issues:

Inspect bunkers after rainfall for water pooling and surface damage.

Use soil probes to assess drainage efficiency.

Re grade bunkers where needed and install supplemental drainage systems.

Schedule regular maintenance to clear drainage channels.

Enhance soil structure with organic amendments and proper turf management.

Final Thoughts

The challenges of turf grass management are multifaceted and interdependent. By developing a robust understanding of fungal diseases, weed dynamics, pest pressures, and the subtleties of bunker drainage, you empower yourself to make proactive and effective management decisions. The integrated strategies provided in this chapter demonstrate that success in turf management does not come from singular solutions, but from the convergence of careful diagnosis, cultural best practices, and targeted chemical or mechanical interventions.

Embrace these strategies as part of a continuous improvement cycle. Document successes and setbacks, adjust protocols based on seasonal data, and remain flexible as conditions change. With these comprehensive solutions and checklists at your fingertips, you are well equipped to tackle the most common challenges faced by golf courses, ensuring that your turf not only survives but thrives year after year.

Chapter 9: Weather Related Challenges

When it comes to maintaining high quality turf on a golf course, weather is one of the most significant external factors that can either make or break your seasonal management strategy. Whether it's the relentless heat and drought of midsummer, an unexpected deluge, or the sudden frost of early winter, these climatic events require rapid adaptation and decisive action. In this chapter, we delve into three interrelated areas that are critical for weather related turf management:

1. Addressing Extreme Heat, Drought, and Excess Rainfall: Learn how to manage the adverse effects of heat stress, water scarcity, and overstimulation due to heavy rainfall by using precision irrigation, cultural practices, and advanced water conservation techniques.

2. Emergency Irrigation Repairs: Discover protocols for quickly diagnosing and repairing critical irrigation system failures triggered by weather extremes, ensuring your turf receives appropriate hydration even during emergencies.

3. Responding to Frost Damage Concerns: Understand the impact of frost on turf health and learn proactive measures and recovery strategies to mitigate frost damage while safeguarding the integrity of your greens, fairways, and tees.

By developing an integrated weather response plan, superintendents can ensure that their courses not only weather extreme conditions but also emerge more resilient and ready for the next season.

Addressing Extreme Heat, Drought, and Excess Rainfall

Understanding the Impacts of Weather Extremes

Extreme weather conditions be they soaring temperatures, prolonged drought conditions, or waves of excess rainfall create unique challenges for turf grass. Each scenario stresses the turf in different ways:

Extreme Heat: High temperatures lead to rapid evapotranspiration, which can dehydrate the turf before water reaches the roots. This often results in cellular damage, scorching, and a decline in turf density.

Drought Conditions: Extended periods without adequate rainfall force the turf to rely on stored moisture. Without deep, consistent watering, the turf will produce shallow roots, making it extremely vulnerable to water deficits.

Excess Rainfall: Heavy or prolonged rainfall can flood areas of the course, causing waterlogging that leads to oxygen deprivation in the root zone. Additionally, excess moisture can promote fungal diseases and lead to rapid nutrient leaching from the soil.

Strategies for Managing Extreme Heat and Drought

A proactive approach to managing water supply and heat stress is essential. Consider the following strategies:

A. Precision Irrigation and Smart Controllers

Utilize Smart Irrigation Systems: Modern irrigation controllers that incorporate real time weather data (temperature, humidity, wind speed, and evapotranspiration rates) can adjust watering schedules on the fly. This ensures that watering is neither excessive during heavy rains nor insufficient during drought conditions.

Zonal Irrigation: Divide the course into various irrigation zones based on sun exposure, soil composition, and turf grass type. Tailoring water application to the unique needs of each zone helps conserve water and ensures that moisture reaches deeper into the root zone.

B. Deep Watering Techniques

Early Morning Watering: Watering early in the morning reduces evaporation losses and gives the turf a chance to absorb moisture before the heat of the day. This approach promotes the development of a deeper, more resilient root system.

Cycle and Soak Principles: Instead of delivering one long watering session, subdivide the water allocation into several shorter sessions. This "cycle and soak" approach helps water penetrate more deeply and prevents runoff on slopes or compacted soils.

C. Cultural and Soil Management

Soil Amendments: Incorporate organic matter or wetting agents that help improve soil structure and water retention. Organic amendments not only enhance moisture holding capacity but also improve nutrient availability.

Drought Tolerant Turf Varieties: If re sodding or over seeding is part of your longer term strategy, consider planting turf grass varieties that are well adapted to heat and drought conditions. These cultivars often require less water and recover more quickly from heat stress.

Managing Excess Rainfall

Excess rainfall, while seemingly beneficial, presents issues that require careful management:

A. Enhancing Drainage

Improve Surface Drainage: Ensure that all playing areas and utility zones are graded with appropriate slopes to channel water away from critical turf areas. In areas where water pooling is evident, consider installing additional drainage infrastructure such as French drains or subsurface drainage pipes.

Maintain Drainage Equipment: Regularly inspect and clear all drainage systems catch basins, ditches, and underground pipes to prevent blockages that lead to localized flooding.

B. Adjusting Irrigation Schedules

Suspend Irrigation as Necessary: When significant rainfall is forecast or has recently occurred, temporarily suspend scheduled irrigation. This avoids overwatering and minimizes the risk of disease and nutrient leaching.

Monitor Soil Moisture: Use soil moisture sensors to continuously track water levels. This real time feedback can help you decide precisely when to resume irrigation.

Real World Example

At a coastal golf course known for its erratic weather, the superintendent installed a network of smart controllers and soil moisture sensors that provided real time feedback on moisture levels. During a summer heatwave, the system automatically increased watering intervals, while during a torrential rain event, the irrigation schedule was suspended entirely. As a result, the course maintained consistent turf health with minimal water waste even in the face of extreme conditions.

Diagram Idea

Create a multi level flowchart that illustrates the decision making process for weather extremes. The chart might start with "Weather Data (Temperature, Humidity, Rainfall)" and branch into:

Extreme Heat/Drought Adjust irrigation schedule (deep, early morning watering, cycle and soak)

Excess Rainfall Suspend irrigation, check drainage, and monitor soil moisture.

Checklist Managing Extreme Heat, Drought, and Excess Rainfall

Install and calibrate smart irrigation controllers with weather integration.

Divide the course into distinct irrigation zones for customized watering.

Schedule early morning watering sessions; adopt cycle and soak practices.

Integrate soil amendments and consider drought tolerant turf grass cultivars.

Regularly inspect and improve surface and subsurface drainage.

Temporarily suspend irrigation during or after heavy rainfall.

Monitor soil moisture continuously and adjust schedules accordingly.

Document water usage and weather patterns to fine tune future protocols.

Emergency Irrigation Repairs

The Impact of Weather on Irrigation Systems

Irrigation systems are the lifelines for turf, especially during periods of drought or extreme heat. However, these systems themselves are vulnerable to weather induced stresses. Heavy rainfall can cause physical damage to components, while high temperatures and prolonged sunlight may degrade hoses, pipes, and sprinkler heads over time. The ability to respond quickly to irrigation system failures is vital to prevent further turf stress.

Diagnosing Common Irrigation System Failures

A. Visual Inspections

Leaks and Breaks: Inspect the entire system for signs of water leaks, broken pipes, or dislodged sprinkler heads. After heavy rainfall, check for standing water near irrigation lines that may indicate system failure.

Pressure Drops: Monitor pressure gauges in various zones. A significant drop in pressure can indicate a blockage, ruptured pipe, or malfunctioning pump.

B. Sensor and Controller Diagnostics

Automated Alerts: Many modern systems come equipped with sensors that can provide alerts for unusual pressure or flow deviations. Responding to these alerts promptly can minimize the duration of a malfunction.

Manual Checks: Despite high tech systems, a periodic manual walkthrough of the irrigation system is indispensable. Look for obvious wear, sun damage, or misalignment in sprinkler heads and pipes.

Steps for Emergency Irrigation Repairs

A. Immediate Response Protocol

Shut Down Affected Zones: If a leak or break is detected, immediately shut down the affected irrigation zone to limit water loss and prevent further damage.

Temporary Fixes: Apply quick fixes such as using repair tape or clamps to contain minor leaks until permanent repairs can be made. Ensure that safety protocols are followed, especially when working with pressurized systems.

B. Coordinated Repair Efforts

Mobilize Maintenance Teams: Assign trained personnel to inspect, diagnose, and repair the malfunction. In cases where specialized parts or tools are required, have an emergency repair kit on hand.

Document the Repair: Record the details of the malfunction location, cause, and repair actions. This documentation helps in adjusting future maintenance schedules and budgeting for replacements.

C. Test and Recalibrate

Post Repair Testing: Once repairs are completed, conduct a full test of the system for proper functioning. Check pressure levels and examine all previously affected areas for recurrence.

Integrate Feedback: Use the incident as a learning experience. Adjust preventive maintenance schedules to target vulnerabilities revealed during the emergency.

Real World Example

Following an unexpected deluge that caused several irrigation lines to rupture, a superintendent quickly divided the course into isolated zones and implemented temporary repairs using repair clamps and duct tape. The swift action minimized water waste and turf stress, and a follow up inspection scheduled during the off peak hours ensured that permanent repairs were executed before the system was fully reactivated. The detailed documentation of the incident later informed an upgrade plan for the aging underground piping.

Diagram Idea

Develop a troubleshooting flowchart that begins with "Irrigation Malfunction Detected" and branches into:

Visual Inspection, Identify Leak/Blockage, Shut Down Zone, Implement Temporary Fix, Schedule Permanent Repair.

Automated Sensor Alert, Manual Verification, Immediate Shutdown (if necessary), Repair Process.

Checklist Emergency Irrigation Repairs

Regularly inspect irrigation lines, sprinkler heads, and pressure gauges.

Install sensors and configure automated alerts for pressure/flow issues.

Maintain an emergency repair kit with essential parts (tape, clamps, repair sleeves).

Immediately shut down malfunctioning irrigation zones upon detection.

Execute temporary repairs promptly and schedule permanent repairs as soon as possible.

Test and recalibrate the system post repair.

Document the incident and update maintenance protocols accordingly.

3. Responding to Frost Damage Concerns

Understanding Frost Damage and Its Effects

Frost damage can occur when temperatures drop sharply, often overnight or during early spring, resulting in the formation of ice crystals within plant cells. This damage can lead to discoloration, tissue rupture, and even death of sensitive turf grass species. Frost events are particularly problematic when followed by fluctuating temperatures that repeatedly thicken and melt ice, further stressing the turf.

Identifying Frost Damage in Turf

A. Visual Symptoms

Discoloration: Frost damaged turf may turn a pale, grayish white or even exhibit brown patches where cell structures have been severely compromised.

Crisping and Blighting: Beneath a frost event, leaf tips and edges may become crisp or brittle, indicating cellular dehydration and rupture.

Uneven Recovery: In some cases where frost damage is localized, certain areas may recover more slowly than surrounding healthy turf often visible in patchy regrowth during early morning inspections.

B. Timing and Environmental Indicators

Early Warning Signs: Monitor weather forecasts vigilantly. Frost alerts, even during the transition periods of early spring or late fall, should trigger preemptive action.

Microclimates: Be aware of low lying areas or zones with poor air circulation on the course. These microclimates are more susceptible to frost accumulation.

Preventative Strategies for Frost Damage

A. Cultural Practices

Adjust Mowing Practices: Prior to forecasted frost events, refrain from cutting the turf too short. Taller grass can offer protective insulation for the roots and provides a microclimate that retains heat.

Irrigation Adjustments: Paradoxically, applying a light, controlled amount of water before expected frost can create a protective layer of ice on the turf blades that, if managed correctly, prevents deep cellular damage. However, this technique must be balanced carefully to avoid over saturation.

Delayed Fertilization: Avoid heavy nitrogen applications immediately before the frost season as lush, tender growth is especially vulnerable to frost injury. Instead, focus on balanced, gradual fertilization that encourages toughness and resilience.

B. Mechanical and Chemical Aids

Anti Transpirants: The use of anti transpirant sprays can reduce water loss from the turf during periods of freezing temperatures. These products form a thin protective film on leaf surfaces.

Physical Frost Protection: In some high value or high sensitivity areas like greens, consider temporary protective covers or frost blankets if a severe frost is anticipated. While not practical for the entire course, selective application can save critical turf.

C. Post Frost Recovery Tactics

Rapid Response: Once temperatures begin to rise, immediately assess the extent of frost damage. Avoid further stress by delaying mowing until the turf shows signs of recovery.

Nutrient Recovery: Apply a recovery fertilizer with a balanced nutrient profile to support cell repair and new growth once the frost threat has passed.

Irrigation Management: Adjust irrigation to gently rehydrate damaged turf without flooding it. Light, frequent watering that avoids sudden saturation will help the recovery process.

Real World Example

After an unexpected frost broke several nights early in the season, a superintendent monitored the greens closely and delayed their first mowing until clear signs of recovery were visible. In addition, a light anti transpirant was applied to vulnerable areas, and a recovery fertilizer, low in nitrogen but rich in potassium and phosphorus, was broadcast over the course. These measures collectively resulted in a noticeably improved recovery rate, with minimal patchiness noted during the subsequent warm up period.

Diagram Idea

A diagram or series of images can illustrate the progression of frost damage:

1. Normal turf condition before frost.

2. Formation of frost and initial damage (discoloration, crisp edges).

3. Protective measures (anti transpirant application, modified irrigation).

4. Post frost recovery, highlighting improved turf condition after management interventions.

Checklist Responding to Frost Damage Concerns

Monitor weather forecasts daily during transitional seasons.

Identify microclimates within the course that are at higher risk for frost accumulation.

Avoid mowing turf excessively short before frost events.

Consider a light pre frost irrigation to form a protective ice layer, if appropriate.

Apply anti transpirants in sensitive areas when frost is forecast.

Delay mowing and heavy fertilization until after the frost threat has passed.

Use recovery fertilizers and light irrigation post frost to promote healing.

Document areas of frost damage and assess the effectiveness of protective measures.

Conclusion: An Integrated Approach to Weather Challenges

Weather related challenges whether they manifest as extreme heat, unexpected drought, surges of heavy rainfall, emergency irrigation breakdowns, or damaging frost are among the most dynamic issues facing golf course superintendents. However, by adopting an integrated approach that combines proactive cultural practices, advanced technology, and rapid response procedures, you can transform these challenges into manageable components of your overall turf management program.

Key points include:

For Extreme Heat, Drought, and Excess Rainfall: Precision irrigation, zonal management, and proper soil and turf maintenance provide a solid foundation to counter water related stresses.

For Emergency Irrigation Repairs: Quick identification, immediate zone shutdowns, temporary repair measures, followed by permanent solutions, ensure that the irrigation system remains a reliable asset regardless of sudden weather disruptions.

For Frost Damage: Preventative measures such as adjusted mowing practices, pre and post frost irrigation management, anti transpirants, and recovery protocols help shield the turf from freezing injury while promoting swift recovery.

By understanding the nuances of how weather extremes interact with turf growth and infrastructure, you can anticipate potential problems and take steps that preserve both the quality of the turf and the longevity of critical assets. Documenting each intervention and learning from real world outcomes builds an evolving playbook that continuously enhances your operational resilience. In this way, even when the weather is unpredictable, your course can remain a model of excellence and reliability year after year.

Final Summary Checklist for Weather Related Challenges

Managing Extreme Heat, Drought, and Excess Rainfall:

Implement smart, zonal irrigation systems integrated with real time weather data.

Use early morning, deep watering practices and cycle/soak techniques.

Enhance soil water retention with organic amendments and wetting agents.

Improve grading and drainage to handle excess rainfall.

Suspend irrigation during heavy rainfall events and continuously monitor soil moisture.

Emergency Irrigation Repairs:

Establish routine visual and sensor based inspections.

Develop an emergency shutdown protocol for malfunctioning zones.

Maintain an emergency repair kit and detailed log of repairs.

Test the system thoroughly post repair and recalibrate as necessary.

Responding to Frost Damage:

Monitor weather forecasts and identify at risk microclimates.

Adjust mowing heights and postpone heavy fertilization before frost.

Use anti transpirants and, if feasible, apply protective coverings in sensitive areas.

Initiate cautious recovery practices with post frost fertilizers and gentle irrigation.

Document frost damage areas and assess protection measure effectiveness.

Final Thoughts

Weather related challenges are an ever present part of golf course management. By integrating precision technology with proven cultural practices, you create a proactive system that not only responds to but preempts the adverse effects of extreme weather events. Whether it's combating the fierce pressures of a drought or mitigating the sudden impacts of a frost, the strategies outlined in this chapter equip you with a robust toolkit for weathering any storm.

In the face of nature's unpredictability, a prepared and flexible approach is your greatest asset. Embrace continuous improvement by documenting outcomes, refining procedures, and staying informed on the latest technological and operational innovations. With these measures in place, your course will not only survive weather extremes but thrive, maintaining exceptional playability and aesthetic appeal through every season.

Chapter 10: Sustainability & Environmental Stewardship

In today's competitive landscape, golf courses are under increasing scrutiny to not only provide an exceptional playing environment but also to do so using environmentally responsible methods. As water scarcity, chemical runoff, and energy usage become critical concerns, sustainable turf management and environmental stewardship are no longer optional they are central to long term operational success and community goodwill. This chapter outlines comprehensive strategies in three key areas:

1. Sustainable Turf Management Practices

2. Water Conservation and Efficiency Planning

3. Eco Friendly Fertilization and Chemical Applications

Each section is designed to guide you through actionable methodologies that reduce environmental impact while ensuring that our turf remains healthy, resilient, and visually appealing.

Sustainable Turf Management Practices

A. Defining Sustainability on the Golf Course

Sustainable turf management goes beyond traditional maintenance by balancing aesthetic appeal, functional performance, and environmental responsibility. This approach seeks to reduce the lifecycle impacts of turf maintenance while enhancing biodiversity and preserving resources. Key aspects include:

Integrated Pest Management (IPM): Using cultural, biological, and minimal chemical controls to reduce pesticide usage.

Cultural Practices: Prioritizing strategies such as proper mowing, balanced fertilization, and aeration that reduce stress on the turf.

Biodiversity: Promoting a diverse turf ecosystem which not only increases resilience against pests and diseases but also supports beneficial insects and soil microbes.

B. Cultural Practices for Sustainable Turf

Optimized Mowing and Aeration Maintaining proper mowing heights reduces canopy density; this improves airflow, reduces moisture retention, and naturally suppresses fungal diseases. Coupled with regular aeration, these practices help relieve soil compaction and encourage deeper root systems, which are less reliant on surface water.

Balanced Fertilization Programs Using balanced, slow release fertilizers ensures that nutrient release coincides with the turf's absorption needs. Moreover, shifting from high input synthetic fertilization to organic amendments can maintain soil fertility over the long term and reduce harmful runoff.

Integrated Pest Management (IPM) IPM involves monitoring pest populations, using biological controls (such as beneficial predators or microbial inoculants), and, if necessary, applying chemical treatments on a targeted, minimal basis. This strategy minimizes ecosystem disruption and reduces harmful residue levels.

C. Promoting Biodiversity and Ecological Health

Native Plant Integration: Where possible, integrating native grasses or wildflower strikes around the course periphery can promote biodiversity and provide habitat for beneficial organisms.

Reduced Tillage and Minimal Disturbance: Minimizing soil disturbance through careful planning of mechanical operations protects soil structure and its living constituents.

Habitat Creation: Creating buffer zones or wildlife corridors can enhance ecological diversity, helping to support pollinators, insectivorous birds, and other beneficial wildlife that contribute to pest control naturally.

Real World Example

At a progressive golf club in the Pacific Northwest, the superintendent implemented an integrated sustainability program. By adjusting mowing schedules to maintain a slightly higher grass height and transitioning to organic amendments during fertilization, the course recorded a 30% reduction in chemical inputs. Subsequent monitoring showed improved soil organic matter and increased beneficial insect populations, which in turn reduced pest outbreaks.

Diagram Idea

A layered diagram might illustrate "The Sustainable Turf Management Cycle." The outer layer shows cultural practices (mowing, aeration, fertilization) while the inner layers depict biodiversity enhancements and IPM measures. Arrows can indicate the beneficial feedback loops less chemical use leading to healthier soil, which supports natural pest suppression.

Checklist Sustainable Turf Management Essentials

Establish a routine for mowing that maintains optimal height and encourages deep rooting.

Schedule regular aeration to alleviate soil compaction.

Transition to balanced, slow release fertilization and incorporate organic amendments.

Develop and implement an integrated pest management (IPM) plan.

Promote biodiversity by integrating native species and creating buffer zones.

Record the outcomes of sustainable practices to monitor improvements and adjust protocols over time.

Water Conservation and Efficiency Planning

A. The Importance of Water Stewardship

Water is a critical and increasingly limited resource. Golf courses are significant water users, so optimizing water management practices is essential both for sustainability and for controlling operational costs. A robust water conservation plan not only minimizes usage but also enhances water quality through reduced runoff and improved soil retention.

B. Implementing Smart Irrigation Systems

Digital Controllers and Sensors Smart irrigation controllers use real time weather data (temperature, humidity, and wind speed, rainfall) to adjust watering schedules accordingly. Soil moisture sensors installed in key areas provide continuous feedback, ensuring that water is applied only when needed.

Zonal Irrigation Practices Dividing the course into irrigation zones based on soil type, sun exposure, and turf variety allows for targeted watering. This segmented approach prevents overwatering and promotes deep root growth by tailoring water application to actual needs.

Cycle and Soak Techniques Instead of a single, prolonged watering session, or "cycle and soak" involves dividing water delivery into several shorter cycles. This method reduces runoff, allows water to infiltrate deeper into the soil, and avoids saturation that may lead to disease.

C. Soil Management for Improved Water Retention

Organic Matter and Amendments Regularly adding organic matter such as compost, biochar, or other soil amendments can improve soil structure and water holding capacity. Healthier soils retain moisture longer, reduce the frequency of irrigation, and support a robust root system.

Proper pH Management Maintaining an optimal pH (typically between 6.0 and 7.0 for most turf grasses) ensures nutrient availability and optimal water absorption. To adjust pH, liming materials or sulfur applications may be used as recommended by soil tests.

Use of Wetting Agents In soils that develop hydrophobic layers, applying wetting agents can promote uniform water distribution. These agents help water penetrate deeply, ensuring moisture efficiently reaches the root zone.

D. Long Term Water Efficiency Initiatives

Develop a Water Use Budget: Establish a baseline water usage level and set annual reduction goals. Monitor use via smart irrigation systems and adjust practices based on seasonal performance.

Public Outreach and Transparency: Sharing water efficiency goals and progress with stakeholders and the local community not only enhances reputation but also fosters collaboration and learning among peers.

Real World Example

A golf course in a water scarce region integrated smart controllers, zoning, and cycle soak irrigation techniques to optimize their water use substantially. Their efforts resulted in a 35% reduction in water usage over three seasons without compromising turf quality. Additionally, the use of organic amendments improved the soil's water retention, allowing the course to sustain a healthy playing surface even during short term drought periods.

Diagram Idea

A decision tree or flowchart depicting the water efficiency planning process could include:

Starting with weather and soil data collection,

Adjusting irrigation zones,

Implementing cycle and soak watering methods,

Continuous monitoring via sensors,

Ending with feedback loops for future improvements.

Checklist Water Conservation and Efficiency Planning Essentials

Install and calibrate soil moisture sensors across key zones.

Deploy smart irrigation controllers integrating real time weather data.

Segment the course into distinct irrigation zones based on environmental factors.

Utilize cycle and soak watering techniques to maximize infiltration.

Regularly add organic matter and amend soil as necessary for enhanced water retention.

Monitor and record water usage to track efficiency improvements.

Adjust irrigation schedules based on seasonal and weather related factors.

__ Communicate water conservation goals and progress to stakeholders.

Eco Friendly Fertilization and Chemical Applications

A. The Need for Eco Friendly Inputs

Eco friendly fertilization and chemical applications address two critical concerns: reducing environmental pollution and safeguarding human, animal, and plant health. Traditional chemical fertilizers and pesticides when misapplied can lead to soil degradation, water contamination, and adverse ecosystem impacts. Shifting to eco friendly, sustainable practices helps create a resilient turf system that minimizes negative externalities.

B. Sustainable Fertilization Practices

Organic Fertilizers and Soil Amendments
Using organic fertilizers sourced from manure, compost, and other natural materials enriches soil organic matter and supports microbial health. These inputs tend to release nutrients slowly, reducing the risks of leaching and nutrient runoff.

Slow Release and Controlled Release Fertilizers Modern technologies have enabled the development of slow release fertilizers that synchronize nutrient release with plant uptake. This not only maximizes nutrient use efficiency but also reduces the environmental load of surplus nutrients.

Precision Application Techniques Employ technologies such as GPS guided spreaders and variable rate applicators to ensure that fertilizers are applied uniformly at appropriate rates. This precision reduces waste, minimizes the risk of over application, and protects sensitive areas from excessive nutrient buildup.

C. Eco Friendly Pest and Chemical Controls

Biological Pesticides and Biofungicides These products involve naturally derived substances or live beneficial organisms that suppress pests and diseases without the harmful residues associated with conventional chemicals. For example, microbial biocontrol agents can target specific pathogens while leaving other organisms unharmed.

Integrated Chemical Application Strategies
When chemical inputs are necessary, adopt an

integrated approach that rotates product classes to prevent resistance, focuses on spot treatments rather than broad applications, and relies on products with lower toxicity profiles. This minimizes long term environmental hazards and preserves safer ecosystems.

Formulation Innovations for Reduced Environmental Impact Advances in nanotechnology and formulation science have led to the development of products that require lower dosages while maintaining efficacy. These innovative solutions reduce the total chemical load on the environment and decrease the frequency of applications.

D. Certification and Labeling Programs

Eco Certification: Look for products certified by recognized sustainability agencies. These products have undergone rigorous testing to ensure that they have lower environmental impacts.

Transparency and Traceability: Suppliers that provide full transparency in ingredient sourcing, production methods, and environmental impact assessments can be trusted partners in your sustainability journey.

Real World Example

A renowned golf course in California transitioned 60% of its fertilizer program from synthetic to organic and slow release products, while also integrating biological pest controls where feasible. Over several seasons, the course experienced improved soil health, lower incidence of pest outbreaks, and markedly reduced chemical runoff into adjacent water bodies. Furthermore, the transition enhanced the course's reputation with environmentally conscious members and local community stakeholders.

Diagram Idea

Consider creating a side by side comparison chart that details traditional chemical inputs versus eco friendly alternatives. The chart can list parameters such as nutrient release rates, application frequency, environmental impact, and cost efficiency. This visual tool can succinctly demonstrate the benefits of transitioning to sustainable inputs.

Checklist Eco Friendly Fertilization & Chemical Application Essentials

Evaluate current fertilizer and pesticide programs against sustainable alternatives.

Transition to organic and slow release fertilizers wherever possible.

Implement precision application methods using GPS and variable rate technology.

Incorporate biological pest control agents and biofungicides in an integrated pest management (IPM) framework.

Rotate chemical classes to prevent resistance.

Source products with trusted eco certifications and environmental labels.

Continuously track and adjust applications based on soil and turf response.

Educate staff on the benefits and proper use of eco friendly products.

Conclusion: Toward a Greener Future in Turf Management

As environmental concerns become ever more paramount, sustainable and eco friendly practices in golf course management are not only beneficial they're essential. By embracing sustainable turf management strategies, implementing robust water conservation plans, and transitioning to eco friendly fertilization and chemical applications, superintendents can significantly reduce the ecological footprint of their courses while maintaining or even enhancing turf quality.

This chapter has provided a holistic view of sustainability and environmental stewardship by:

Emphasizing cultural and ecological practices that support a resilient, diverse turf ecosystem.

Outlining water conservation strategies that leverage smart technology and soil management techniques to optimize water use.

Presenting innovative, eco friendly alternatives for fertilization and chemical applications that minimize adverse environmental impacts.

The integrated approach presented here is not an endpoint but an evolving process of continuous improvement. By documenting outcomes, monitoring key performance indicators, and staying informed about the latest technological advances and industry best practices, you can create a sustainable turf management system that meets both present demands and future challenges.

Above all, true sustainability in turf management is about striking a balance enhancing the beauty and playability of the course while serving as a custodian of natural resources. With each step you take toward eco friendlier methods, you contribute to a cleaner environment, healthier ecosystems, and a more socially responsible legacy.

Final Summary Checklist for Sustainability & Environmental Stewardship

Sustainable Turf Management Practices:

Implement integrated cultural practices (optimized mowing, proper aeration, balanced fertilization).

Adopt integrated pest management (IPM) with biological controls.

Promote biodiversity and incorporate native species to support ecological health.

Document performance metrics and refine practices accordingly.

Water Conservation and Efficiency Planning:

Install smart irrigation systems with real time weather and soil moisture monitoring.

Segment the course into irrigation zones and implement cycle soak watering techniques.

Enhance soil structure with organic amendments and wetting agents.

Establish and monitor a water use budget, adjusting protocols as needed.

Eco Friendly Fertilization and Chemical Applications:

Transition to organic and slow release fertilizers.

Utilize precision application techniques to minimize wastage.

Incorporate biological pest control products within an IPM framework.

Source products with trusted eco certifications and ensure proper rotation of chemicals.

Continually evaluate and adjust chemical inputs based on turf performance.

Final Thoughts

The pursuit of sustainability in golf course management is a journey one that requires commitment, innovation, and an unwavering dedication to environmental stewardship. Whether it's refining cultural practices, harnessing the power of smart technology, or embracing eco friendly chemical alternatives, every step you take drives you further toward a greener future.

In the interplay between resource efficiency and operational excellence, sustainability is the ultimate convergence of financial prudence and environmental responsibility. By adopting the integrated strategies outlined in this chapter, you can not only reduce the ecological footprint of your course but also cultivate a legacy of stewardship that will inspire fellow professionals and resonate with your community.

As we conclude this chapter, remember that sustainable practices are dynamic, and continuous learning is key. Observe outcomes, solicit feedback, and always be prepared to refine your approaches based on new research and technological advancements. In doing so, you ensure that your course remains a model of excellence both in playability and in its contribution to a healthier, more sustainable environment.

Technology & Innovation

Chapter 11: Precision Turf Management

Introduction

In today's competitive and environmentally conscious world, precision turf management is transforming how golf courses are maintained. Gone are the days when decisions were based solely on visual observation and historical trends. Now, advanced technologies such as high resolution GPS mapping, artificial intelligence (AI) algorithms that analyze multispectral imagery, smart irrigation controllers, and real time soil sensor networks provide superintendents with actionable insights. These technologies not only help achieve more uniform and resilient turf but also optimize resource use by ensuring that inputs such as water, fertilizers, and pesticides are applied in the right amounts, at the right time, and only where needed.

This chapter is divided into three primary sections:

GPS and AI for Turf Analysis – Leveraging remote sensing and machine learning for precise turf assessment.

Smart Irrigation Techniques – Using technology driven systems for optimal watering.

Soil Sensor Technology for Efficiency – Integrating wireless sensor data to drive decisions on soil conditions.

By integrating these precision management strategies, you can shift your turf maintenance practices from reactive to proactive, reducing waste while maximizing performance and sustainability.

GPS and AI for Turf Analysis

A. Harnessing the Power of Precise Data Collection

Advanced GPS technology combined with AI is revolutionizing turf analysis. Superintendents can now acquire detailed, high resolution maps of the entire course, drawing on data from drones equipped with multispectral cameras or satellite imagery. These images are georeferenced via GPS, allowing for zone specific analysis. With this information, you can:

Identify areas of turf stress, such as patches suffering from nutrient deficiency, disease, or insect damage.

Map differences in growth, density, or color across various zones (for example, comparing the greens, tees, fairways, and roughs).

Correlate these physical variations with underlying soil conditions, irrigation patterns, or historical maintenance practices.

B. AI Powered Analysis and Predictive Insights

Artificial Intelligence (AI) takes this a step further by processing massive datasets to predict potential issues before they become highly visible. Machine learning algorithms trained on historical turf performance data, spectral signatures, and field conditions can provide:

Predictive Alerts: AI can forecast areas at risk of drought stress, fungal disease outbreaks, or pest infestations by recognizing early signs and correlating them with past events.

Customized Treatment Recommendations: Detailed maps suggesting where additional nutrients or water might be needed, or where pest control interventions are advisable.

Improved Calibration of Equipment: Helping calibrate sprayers, mowers, and irrigation systems to ensure that every square meter of turf receives precisely the right treatment.

Real World Example: A leading golf course in the Pacific Northwest integrated drone generated multispectral imagery with an AI driven analysis tool. Within a few weeks, the system identified subtle regional variations in turf health that were invisible to the naked eye. Based on these insights, the superintendent adjusted nutrient applications and irrigation schedules for specific zones, ultimately reducing chemical usage by 20% while improving turf uniformity by over 25%.

C. Visualization and Decision Making Tools

To facilitate rapid decision making, the AI software typically presents data using interactive maps and dashboards. These visualizations allow you to:

Compare real time turf conditions with historical baselines.

Identify hotspots where immediate intervention is required.

Coordinate with maintenance teams to plan precise, localized treatments.

Diagram Idea: Develop a layered flow diagram that begins with data acquisition via drones and sensors, moves to AI based analysis that produces a detailed turf health map, and culminates with actionable insights (e.g., "Zone A needs additional nitrogen" or "Zone B requires increased irrigation"). Arrows can illustrate the data flow and the feedback loop for adjustments.

Checklist GPS and AI for Turf Analysis

Deploy drones or access high resolution satellite imagery to capture multispectral data.

Ensure all imagery is accurately georeferenced using modern GPS technology.

Integrate collected data into an AI platform trained on historical turf performance.

Regularly generate and review turf health maps to identify stress zones.

Use AI generated treatment recommendations to adjust irrigation, fertilization, and pest control practices.

Document changes and compare before and after results to evaluate success.

Smart Irrigation Techniques

A. The Role of Precision Irrigation

Water management is at the heart of turf maintenance, and smart irrigation is essential for efficiency, especially under increasingly unpredictable weather conditions. Smart irrigation combines advanced control systems with real time sensor data to deliver water exactly where and when it's needed.

B. Components of a Smart Irrigation System

Digital Irrigation Controllers and Weather Integration:

Smart Controllers: Utilize algorithms that adjust watering schedules based on local weather forecasts, temperature variations, humidity, and evapotranspiration (ET) rates.

Real Time Adjustments: These controllers can modify watering durations, frequencies, and even the pressure to ensure precise water delivery even during rapidly changing conditions.

Zonal Irrigation:

Customized Watering: By dividing the course into distinct zones based on soil type, sun exposure, and turf grass variety, the system tailors water distribution to the specific needs of each area. For instance, sunnier fairways may require more frequent, deeper watering compared to the shaded greens.

Reduction in Water Wastage: Targeted watering prevents over irrigation in some areas while addressing deficiencies in others.

Cycle and Soak Techniques:

Deep Infiltration: Instead of one continuous watering session, cycle and soak watering divides water delivery into several shorter bursts. This method allows water to penetrate deeper into the soil profile, promoting robust root growth.

Minimizing Runoff: Shorter cycles reduce the risk of runoff, particularly in areas with slopes or compacted soils.

C. Integration with Smart Sensors

Soil moisture sensors are critical for smart irrigation. These devices provide continuous, real time feedback on moisture levels deep in the soil, enabling the system to determine if and when watering is required.

Automation: When soil moisture falls below a set threshold, the system activates water delivery; once optimal moisture is reached, it automatically shuts off the water.

Data Recording: Historical sensor data can show trends in water usage and help refine irrigation schedules seasonally.

Real World Example: A golf course in a semi arid region implemented a smart irrigation system that combined digital controllers, zonal management, and soil moisture sensor inputs. Within one season, the course achieved up to a 35% reduction in water usage. The system's cycle and soak method not only improved water infiltration and root depth but also significantly reduced the incidence of drought induced turf stress during extended dry spells.

D. Visualization and Monitoring

Operators can access an integrated dashboard that displays real time water application data, moisture levels for each zone, and historical usage trends. These tools help you make strategic modifications and justify water conservation measures to stakeholders.

Diagram Idea: Create a flowchart that demonstrates the smart irrigation loop:

Sensors capture current soil moisture levels.

Data is transmitted to the smart controller.

The controller adjusts the watering schedule based on real time weather data.

Water is delivered via zonal irrigation systems.

New sensor readings confirm water uptake, completing the loop.

Checklist Smart Irrigation Techniques

Install smart irrigation controllers with real time weather integration.

Divide the course into well defined irrigation zones.

Implement cycle and soak watering techniques to promote deep infiltration.

Deploy soil moisture sensors in critical areas for continuous monitoring.

Adjust irrigation schedules based on sensor feedback and weather forecasts.

Maintain a digital dashboard to track water usage and system performance.

Periodically review sensor and system data to fine tune water budgets.

Soil Sensor Technology for Efficiency

A. Transforming Soil Management

Accurate soil information is the cornerstone of precision turf management. Soil sensor technology offers the ability to monitor key parameters such as moisture content, temperature, and even salinity in real time. These insights empower superintendents to make informed decisions that optimize fertilization, irrigation, and overall turf health.

B. Types of Soil Sensors and Their Functions

Wireless Soil Moisture Sensors:

Functionality: These sensors are buried below the turf surface and continuously measure moisture levels at various depths. They send data wirelessly to a central controller or dashboard.

Benefits: Rapid alerts when moisture falls below or exceeds the desired threshold enable timely irrigation adjustments, effectively preventing both drought stress and overwatering.

Temperature and Salinity Sensors:

Temperature Monitoring: Soil temperature data can predict turf behavior for example, critical temperature thresholds for seed germination or disease outbreaks.

Salinity Management: Monitoring soil salinity is essential, particularly when using reclaimed water or during periods of heavy fertilization. High salinity can stunt root growth and compromise turf health.

Integrated Sensor Systems:

Networked Systems: Modern systems, such as those offered by Soil Scout or Toro's Turf Guard, integrate multiple sensor types into a unified network with digital monitoring capabilities. These platforms enable real time decision making by correlating soil moisture, temperature, and salinity data.

Long Term Data Recording: Historical sensor data provide insights into seasonal trends and long term soil health, informing adjustments in fertilizer and irrigation regimes.

C. Data Driven Decision Making

Armed with precise soil data, superintendents can:

Optimize Resource Application: Apply fertilizers and water exactly where needed avoiding blanket applications that lead to waste and reduced efficiency.

Prevent Soil Compaction and Degradation: Recognize areas of chronic dryness or oversaturation and re amend them accordingly.

Tailor Cultural Practices: Adjust aeration, over seeding, and other cultural practices based on real time data, ensuring that interventions are perfectly timed to meet the turf's needs.

Real World Example: A renowned championship golf course integrated Soil Scout wireless sensors throughout key areas of the course. Continuous data on moisture and temperature allowed the management team to precisely control irrigation and adjust fertilizer rates. This precision not only lowered overall chemical and water use by approximately 30% but also improved turf uniformity, leading to consistently superior playing conditions during both peak and off peak seasons.

D. Visualization and Integration

Data from soil sensors feeds into centralized dashboards that visualize the status of soil conditions across the entire course. These dashboards allow you to see heat maps, trend graphs, and alerts that indicate when conditions are deviating from optimal ranges.

Diagram Idea: A schematic showing a cross section of the turf with embedded soil sensors could illustrate the layers of data collected (moisture, temperature, salinity) and how this information is transmitted wirelessly to a central dashboard. You can include sample graphs or heat maps demonstrating data trends over time.

Checklist Soil Sensor Technology Essentials

Install wireless soil sensors at various depths and strategic locations.

Ensure sensors are calibrated and integrated with the central data management system.

Monitor real time data on soil moisture, temperature, and salinity.

Use historical data to establish baseline conditions and seasonal trends.

Adjust fertilization and irrigation schedules based on sensor feedback.

Maintain regular maintenance and calibration checks on all sensor equipment.

Utilize visualization tools (heat maps, graphs) to support decision making.

Conclusion: The Future of Precision Turf Management

Advances in GPS, AI, smart irrigation, and soil sensor technology are reshaping the landscape of turf management, turning traditional practices on their head. Precision turf management is about leveraging these tools not only to save on inputs and reduce environmental impact but also to create consistently outstanding playing surfaces that stand up to the challenges of changing climates and rigorous usage.

By integrating high resolution turf analysis with AI, you can identify subtle stress factors and tailor treatments with surgical precision. Smart irrigation techniques ensure that every drop of water is used as efficiently as possible, while real time soil sensor data informs daily adjustments that keep the turf robust and resilient. Together, these technologies form a holistic approach that moves maintenance from reactive, broad spectrum interventions to targeted, data driven strategies.

As you refine your methods and integrate these precision tools into your daily operations, you'll find that the benefits extend beyond immediate improvements in turf quality they also contribute to long term sustainability, cost savings, and a stronger competitive edge in the industry.

Remember, precision turf management is an evolving process. Continuous data collection, analysis, and adaptation are key. By documenting outcomes, regularly updating your protocols, and staying current with the latest technological innovations, you can ensure your course remains at the forefront of excellence and environmental stewardship.

Final Summary Checklist for Precision Turf Management

GPS and AI for Turf Analysis:

Deploy high resolution multispectral imagery via drones or satellites.

Georeference data with modern GPS technology.

Integrate captured data into an AI based analysis platform.

Generate detailed turf health maps and zone specific treatment recommendations.

Regularly update and analyze historical data to refine maintenance practices.

Smart Irrigation Techniques:

Install smart irrigation controllers with weather integration.

Segment the course into tailored irrigation zones.

Apply cycle and soak techniques for deep, efficient watering.

Deploy soil moisture sensors for real time feedback.

Monitor water usage and adjust schedules based on sensor data.

Soil Sensor Technology for Efficiency:

Utilize wireless sensors to continuously monitor moisture, temperature, and salinity.

Integrate sensor data with centralized dashboards for visualization.

Use high resolution data to optimize fertilization and irrigation applications.

Regularly calibrate and maintain sensor equipment to ensure data accuracy.

Leverage historical sensor data to inform future soil management practices.

Final Thoughts

Precision turf management represents the convergence of technology and traditional expertise creating a new paradigm in golf course maintenance. As you embrace GPS, AI, smart irrigation, and soil sensor technologies, you'll not only enhance turf quality and operational efficiency but also pave the way for sustainable practices that benefit both your course and the environment. The future of turf management lies in the ability to say with confidence, "Every drop counts, every blade matters, and every decision is data driven."

By investing in these advanced tools and developing an integrated, proactive management strategy, you position your course for long term success and adaptability. The journey toward truly precision based turf management is ongoing, and by continuously refining your approach, you can ensure that your turf not only meets the demands of today but is also well prepared for the challenges of tomorrow.

Chapter 12: Automation & Future Trends

Introduction

Advances in technology have ushered in a new era for golf course management. What once relied solely on periodic human observations and manual interventions is now evolving into a data driven, automated practice. Modern superintendents increasingly leverage tools that offer real time insights into turf health, optimize resource usage, and reduce labor intensive tasks. This transformation is not only about increasing operational efficiency but also about driving sustainability and reducing environmental impact.

In this chapter, we explore three key domains:

Autonomous Mowers and Robotic Systems: How automated machines are taking over routine turf maintenance.

Turf Monitoring Drones and AI Driven Disease Detection: Harnessing remote sensing and machine learning to identify problems early and tailor interventions.

Innovations Improving Golf Course Sustainability: New technologies and practices that lower the ecological footprint while enhancing efficiency.

By embracing automation and future trends, golf course superintendents can transition from reactive management to proactive, precision based maintenance that adapts to changing conditions and expectations.

Economic Considerations in Embracing Advanced Technologies

While the allure of advanced systems such as autonomous mowers, robotic turf maintenance, and drone based monitoring coupled with AI driven disease detection is undeniable, it's important to recognize that these cutting edge tools come with a sizable price tag. In many cases, the high costs associated with the procurement, integration, and maintenance of these systems mean that they are best suited for high end golf courses with deep capital budgets and the financial flexibility to invest in state of the art technology.

For approximately 90 percent of golf courses, especially those operating with more modest financial resources, the initial investment and ongoing expenses of such advanced automation may prove prohibitive. Smaller or mid range courses often must prioritize budget allocations toward essential operations, making it challenging to justify a leap into technologies that not only require significant upfront capital but also demand specialized staff training and continuous technological updates.

This economic reality creates a dichotomy in the industry. On one hand, elite courses are increasingly positioning themselves at the forefront of innovation by integrating robotics and AI to enhance operational precision, environmental sustainability, and overall playing conditions. On the other hand, the majority of facilities continue to rely on more traditional, manual maintenance methods that, while effective, do not offer the same potential for efficiency gains and resource optimization as these advanced systems.

It is essential to view these technologies as an aspirational benchmark that illustrates the future direction of turf management. For many courses, the adoption of precision tools may be a gradual process, beginning with smaller scale pilot projects or selective implementation in high priority areas. Over time, as the technology matures and costs decrease, it may become more accessible to a broader range of courses. Until then, the transformative potential of automation and AI driven systems remains largely within the realm of high end operations.

This perspective helps set realistic expectations about who will likely be the early adopters of such innovation, while still encouraging all golf course managers to consider how incremental technological upgrades in line with their budgetary constraints can improve management practices over the long term.

Autonomous Mowers and Robotic Systems

A. The Evolution of Turf Maintenance

Autonomous mowers and robotic systems are revolutionizing the way golf courses manage routine tasks like mowing, edging, and even light turf care. These systems operate on pre programmed routes, using high precision GPS navigation and onboard sensors to maintain consistent turf height and quality with minimal human input. This technology not only ensures that mowing is carried out at optimal times often during off peak hours but also reduces labor costs and environmental impacts by shifting away from fuel based machinery.

Key Advantages:

Consistency & Efficiency: Autonomous mowers maintain consistent cutting patterns and heights every day, improving turf uniformity.

Reduced Emissions: Many robotic systems run on electric power, cutting down on fuel consumption and lowering greenhouse gas emissions.

24/7 Operation: These devices can operate round the clock, performing routine tasks at times that minimize disruption to play.

Integration with Other Technologies: Robotic mowers can be linked with GPS mapping systems and even with real time data feeds from soil or turf sensors, allowing for adaptive operations based on current field conditions.

B. Real World Success Stories

Consider the example of a golf course in Finland that integrated an entire fleet of autonomous mowers. With over 30 robotic units working seamlessly during the night, the course observed not only decreased labor costs but also significant improvements in turf uniformity and reduced soil compaction. Similarly, courses in regions facing labor shortages have turned to robotic systems as a reliable, scalable solution to maintain high standards while minimizing human oversight.

C. Visualization and Workflow

Diagram Idea: Imagine a flow diagram that shows:

Route Programming: Mapping the course using high resolution GPS.

Data Integration: Incorporating sensor and real time weather data.

Autonomous Operation: The robotic mower cuts on a pre determined route.

Feedback Loop: Data from operations is sent back to a central system, allowing for performance tracking and schedule adjustments.

D. Checklist Autonomous Mowers and Robotic Systems

Evaluate current labor costs and turf maintenance routines.

Research and select autonomous mower models suitable for your course size and terrain.

Develop a phased integration plan that includes staff training and system calibration.

Integrate the robotic systems with GPS mapping and other sensor technologies.

Set up a central dashboard for monitoring performance, scheduling maintenance, and collecting data.

Regularly review system performance against turf health outcomes and adjust operational parameters accordingly.

Turf Monitoring Drones and AI Driven Disease Detection

A. Enhancing Turf Analysis with Drone Technology

Drones are drastically changing the way superintendents view and manage their courses. Equipped with high resolution cameras and multispectral sensors, these unmanned aerial vehicles capture detailed images of every nook and cranny. By flying regularly over the course, drones provide a bird's eye view that reveals subtle variances in turf color, density, and moisture levels that are often invisible from the ground.

Benefits of Drone Monitoring:

High Resolution Mapping: Generate detailed maps that highlight stress zones, areas of potential disease, and regions that may require adjusted fertilization or irrigation.

Efficiency: Cover vast areas in a short time and reduce the need for time consuming on foot inspections.

Data Collection: Drones can be equipped to gather environmental data (e.g., temperature, humidity) that further refines turf management practices.

B. Leveraging Artificial Intelligence for Early Problem Detection

AI driven analysis systems take drone collected data to the next level. By applying machine learning algorithms to multispectral images, AI platforms can predict areas of emerging disease or water stress. This high tech diagnostic process offers several advantages:

Early Warning System: AI can detect subtle color changes and growth abnormalities, providing early warnings before problems become severe.

Targeted Interventions: By pinpointing exact locations that need attention, treatments can be applied precisely, reducing waste and limiting chemical usage.

Predictive Outcomes: Historical data combined with ongoing inputs allows AI systems to forecast potential outbreaks of disease or pest infestations, enabling preemptive action.

C. Real World Application

At a prominent golf course in the Pacific Northwest, drones equipped with multispectral sensors flew weekly over the course. The data was processed by an AI platform, which generated heat maps pinpointing areas at high risk of fungal infections. This allowed the superintendent to tailor fungicide applications to those exact zones, markedly reducing overall chemical use and improving turf health across the course.

D. Diagram and Workflow

Diagram Idea: A comprehensive diagram might illustrate:

Drone Flight & Data Capture: How drones cover the course and record multispectral images.

Data Processing: AI algorithms analyze the images and highlight problem areas.

Visualization: Results are displayed on an interactive dashboard.

Action: Maintenance plans are adjusted based on the AI generated insights.

E. Checklist Turf Monitoring Drones and AI Driven Disease Detection

Invest in drones with high resolution cameras and multispectral sensors.

Set up a regular flight schedule to collect data throughout the season.

Integrate drone data with an AI platform that is calibrated with historical turf performance.

Develop an interactive dashboard for real time visualization of turf health.

Train staff on interpreting AI driven reports and adjusting cultural practices accordingly.

Document treatment actions and correlate with AI data to refine algorithms over time.

Innovations Improving Golf Course Sustainability

A. Integrating Renewable Energy and Efficient Systems

Sustainability extends beyond water or chemical inputs; it encompasses energy use, waste reduction, and long term ecological stewardship. Emerging technologies are providing innovative ways to make golf courses not only better maintained but also environmentally sustainable.

Key Innovations Include:

Renewable Energy Solutions: Solar panels and wind turbines are being integrated into course operations to power irrigation systems, lighting, and automated equipment.

Energy Efficient Equipment: Upgrading to electric or hybrid machinery reduces fuel consumption and greenhouse gas emissions.

Waste Reduction Technologies: Automated recycling systems and on site composting of turf clippings and organic debris minimize waste and reuse valuable nutrients back into the soil.

B. Smart Infrastructure for Resource Management

Innovations in smart infrastructure allow for comprehensive environmental monitoring and resource tracking. Integrating IoT (Internet of Things) devices into course management systems can provide continuous real time data on energy consumption, water usage, and even air quality. This data is invaluable for:

Monitoring Efficiency: Track and optimize resource use, ensuring the most sustainable practices are in place.

Reducing Carbon Footprint: By minimizing energy waste and optimizing operations, courses can significantly lower their environmental impact.

Enhanced Reporting: These systems support transparent reporting to stakeholders and regulatory bodies, reinforcing the course's commitment to sustainability.

C. Real World Example

A championship golf course in California integrated multiple sustainability innovations over a three year period. They transitioned to a fully electric fleet of maintenance vehicles, installed solar panels to power key infrastructure, and adopted a comprehensive IoT system for monitoring energy and water usage. The result was a 40% reduction in energy costs, improved water efficiency, and a marked enhancement in their sustainability profile a model that attracted both memberships and environmental accolades.

D. Diagram and Visualization

Diagram Idea: A side by side comparison chart or infographic illustrating "Before and After Sustainability Innovations" can be very effective. Include parameters such as energy usage, water consumption, and chemical inputs. Icons representing solar panels, electric equipment, and smart sensors can be used alongside data graphs to show tangible improvements over time.

E. Checklist Innovations for Sustainability

Assess current energy, water, and waste management practices.

Plan for integrating renewable energy sources (e.g., solar, wind) into key operations.

Upgrade existing equipment to energy efficient, electric, or hybrid models.

Implement an IoT based monitoring system for continuous tracking of resource usage.

Explore waste reduction strategies, including recycling and on site composting.

Communicate sustainability benchmarks to stakeholders and adjust based on performance data.

Regularly update sustainability protocols based on emerging technologies and best practices.

Conclusion: Embracing the Future of Turf Management

The future of golf course management is increasingly defined by precision, efficiency, and environmental consciousness. By integrating autonomous mowers and robotic systems, leveraging drone and AI technologies for nuanced turf analysis, and adopting cutting edge innovations aimed at sustainability, superintendents are not merely maintaining courses they are transforming them into dynamic, high performance ecosystems.

Automation and advanced data analytics empower proactive decision making, allowing for targeted interventions that reduce waste and optimize inputs. Simultaneously, innovations in renewable energy and smart infrastructure drive sustainability, ensuring that courses can thrive in a world where environmental stewardship is paramount.

As you look to the future, remember that the journey toward precision turf management is ongoing. Continuous innovations and emerging trends will further redefine operational best practices, and staying ahead means embracing these changes wholeheartedly. Empowered by data, technology, and a commitment to sustainability, your golf course can set a new standard for excellence, resiliency, and responsible stewardship in a rapidly evolving industry.

Final Summary Checklist for Automation & Future Trends

Autonomous Mowers & Robotic Systems:

Research and select appropriate robotic platforms.

Develop a phased integration plan with training and maintenance schedules.

Link autonomous systems with GPS mapping and sensor technologies.

Continuously monitor performance and update operational routes based on data feedback.

Turf Monitoring Drones & AI Driven Disease Detection:

Deploy drones equipped with high resolution, multispectral sensors.

Integrate image data with AI based analysis platforms.

Set up interactive dashboards for real time turf health visualization.

Train staff to translate AI insights into targeted maintenance interventions.

Innovations for Sustainability:

Evaluate and integrate renewable energy sources into course operations.

Upgrade equipment to more energy efficient, electric, or hybrid models.

Implement IoT monitoring systems for continuous resource management.

Adopt precision application technologies and waste reduction strategies.

Report sustainability performance and refine strategies based on data.

Final Thoughts

Automation and future trends are not merely passing fads; they represent a paradigm shift in how we manage, sustain, and elevate the playing conditions of golf courses. With advancements in robotics, precision monitoring, and sustainable innovations, turf management is becoming more efficient, responsive, and environmentally responsible than ever before. By embracing these technologies today, you are not only ensuring a competitive edge and improved operational outcomes but also championing a sustainable future for the sport we love.

The tools and strategies covered in this chapter empower you to transform your maintenance operations, reduce environmental impact, and set the stage for continued excellence in an increasingly data driven world. As technology continues to evolve, your ability to adapt and innovate will be the cornerstone of long term success in golf course management.

Chapter 13: Becoming an Effective Leader

Introduction

Effective leadership is at the heart of every successful operation. In golf course management, the role of the leader extends far beyond scientific turf management and technical maintenance it calls for the ability to motivate and guide teams, manage conflicts gracefully, and ensure that every operational aspect of the course runs like a well oiled machine. As a leader, your vision, communication skills, and decision making prowess directly impact not only the efficiency of your operations but also the satisfaction of golfers and staff alike.

This chapter focuses on three critical dimensions of leadership:

Leading Staff Effectively: Inspiring, guiding, and developing a team with a clear vision and shared values.

Conflict Resolution and Golfer Satisfaction: Addressing conflicts internally and externally while upholding a high level of service and responsiveness to golfer concerns.

Course Operational Efficiency: Creating streamlined processes and systems that enhance overall efficiency and promote a culture of continuous improvement.

Each section presents practical tools, real world examples, diagram ideas, and checklists to help you build your leadership skills and foster an environment that drives excellence.

Leading Staff Effectively

A. The Foundation of Great Leadership

Leading staff effectively begins with establishing trust, setting clear expectations, and creating a work environment where everyone feels valued. A strong leader not only guides but inspires by providing a clear vision, demonstrating integrity, and being approachable.

Key Principles for Effective Leadership:

Clear Communication: Open, honest, and regular communication helps ensure that each team member understands their role, responsibilities, and expectations. This helps to reduce ambiguity and fosters a sense of direction.

Delegation with Trust: Empower your team by delegating tasks according to strengths and interests. Effective delegation involves entrusting responsibility while providing the guidance and support needed for success.

Continuous Training and Development: Invest in your staff through ongoing training opportunities, mentoring, and professional development programs. This builds competence and confidence, leading to improved performance and job satisfaction.

Leading by Example: Exemplify the work ethic, values, and dedication you want to see in your team.

When staff observe your commitment to excellence, they are more likely to mirror those behaviors.

B. Effective Communication Techniques

Effective communication is the critical bridge between strategy and execution. Here are some best practices:

Active Listening: Encourage open dialogue by genuinely listening to your team. Validate their feedback and show empathy, which fosters a cooperative environment.

Regular Meetings: Hold frequent team meetings to review progress, address issues, and celebrate successes. These forums allow for collective brainstorming and reinforce a sense of community.

Transparent Decision Making: Be clear about how decisions are made and involve your team when possible. This transparency builds trust and ensures collective ownership of outcomes.

C. Building Team Cohesion and Motivation

Motivated teams are ones where every member feels an intrinsic part of the organization's success.

Recognition and Incentives: Regularly acknowledge individual and team achievements, whether through formal awards, bonuses, or simple expressions of appreciation.

Fostering a Collaborative Culture: Encourage team building activities and open channels for feedback. Creating an environment where ideas are valued helps staff feel empowered and committed.

Conflict Prevention: Establish clear policies regarding behavior and performance that preemptively address potential issues. A cohesive team is less likely to encounter disruptive conflicts, and when issues arise, they can be managed more swiftly.

D. Real World Example

A superintendent at a well regarded course in the Pacific Northwest implemented weekly briefings and quarterly team building retreats. By setting clear short term goals and providing regular feedback, the team's efficiency and morale improved noticeably. The superintendent's openness to listening sessions where staff could voice concerns resulted in adjustments that boosted both individual satisfaction and overall operational performance.

E. Diagram Idea

A possible diagram could depict a "Leadership Cycle" with the following stages:

1. Vision & Strategy Development

2. Open Communication & Delegation

3. Team Training & Development

4. Feedback & Recognition

5. Evaluation & Continuous Improvement
Arrows would indicate the cyclical nature, reinforcing that effective leadership is an ongoing process.

F. Checklist Leading Staff Effectively

Establish a clear, shared vision and communicate it regularly.

Hold regular team meetings and one on one sessions.

Delegate tasks to match individual strengths and provide necessary support.

Develop a training calendar for ongoing staff development.

Implement recognition programs to celebrate achievements.

Encourage a culture of open feedback and continuous improvement.

Lead by example with integrity, commitment, and transparency.

Conflict Resolution and Golfer Satisfaction

A. Importance of Conflict Resolution

Conflict is inevitable, whether stemming from internal team differences or external golfer complaints. Effective conflict resolution is essential to maintain harmony, drive continuous improvement, and ensure that both staff and golfers feel valued and understood.

Core Goals in Conflict Resolution:

Maintain a Positive Environment: Resolve disputes swiftly so that they do not disrupt play or teamwork.

Enhance Golfer Satisfaction: In a service driven environment, addressing golf related complaints effectively can turn a dissatisfied golfer into a loyal patron.

Build Resilient Teams: Use conflicts as growth opportunities by learning from them and implementing changes that strengthen team dynamics.

B. Strategies for Resolving Internal Conflicts

Proactive Communication and Early Intervention

Open Door Policy: Encourage staff to share concerns before they escalate. This fosters a climate of mutual respect and early problem solving.

Active Listening and Mediation: When conflicts arise, listen without bias and mediate discussions to find collective solutions.

Structured Conflict Resolution Processes

Formal Procedures: Establish and communicate clear procedures for resolving disputes. This might include direct mediation, facilitated group meetings, or involving upper management when necessary.

Follow Up: After resolving a conflict, monitor the situation closely to ensure that the issue does not resurface and that team dynamics have improved.

C. Managing External Conflicts for Golfer Satisfaction

Addressing Golfers' Complaints Effectively

Empower Frontline Staff: Train staff to handle complaints with empathy and immediate corrective actions. Ensuring staff have authority to make on the spot decisions can minimize dissatisfaction.

Transparent Communication: When handling a complaint, explain the steps being taken to address the issue. Transparency goes a long way in reassuring golfers that their experience is important.

Creating a Feedback Loop

Surveys and Feedback Mechanisms: Use regular surveys to gauge golfer satisfaction and identify recurring issues.

Responsive Action: Show golfers that their feedback results in tangible changes, thereby fostering trust and long term loyalty.

D. Real World Example

At a high end course, a superintendent faced recurring complaints about slow play and inconsistent green speeds. By instituting a customer feedback program and holding regular debrief sessions with frontline staff, the management team identified process inefficiencies that were the root cause. Moreover, they trained staff in conflict resolution techniques to promptly handle complaints. As a result, golfer satisfaction scores improved markedly, and the course saw an uptick in repeat memberships.

E. Diagram Idea

A diagram could illustrate a "Conflict Resolution Flowchart":

Identify the Conflict: (Internal dispute or golfer complaint)

Gather Information: (Listen and document details)

Intervene Protocol: (Direct mediation for staff or immediate corrective action for golfer issues)

Follow up and Feedback: (Monitor outcomes and adjust processes) This visual tool guides staff through each step in managing conflicts effectively.

F. Checklist Conflict Resolution and Golfer Satisfaction

Establish clear, formal procedures for both internal and external conflict resolution.

Train staff in active listening, mediation, and customer service.

Implement regular feedback mechanisms (surveys, suggestion boxes, digital platforms).

Ensure prompt and empathetic responses to golfer complaints.

Hold debrief sessions after conflict resolutions to assess outcomes and refine processes.

Document conflicts and resolutions to identify trends and improve future responses.

Driving Course Operational Efficiency

A. The Role of Operational Efficiency in Leadership

Operational efficiency is a critical component of effective leadership. Streamlined operations not only reduce costs and improve service quality but also free up resources and time that can be reinvested in strategic initiatives and staff development.

Key Ingredients for Enhanced Operational Efficiency:

Process Optimization: Identify and reduce redundancies in day to day operations.

Technology Integration: Leverage data driven platforms and automated systems to improve workflow and decision making.

Consistent Standards: Establish standard operating procedures (SOPs) to ensure uniformity and predictability in operations.

B. Enhancing Efficiency Through Innovation

Embracing Automation and Technology

Data Analytics: Use performance data to identify bottlenecks and opportunities for improvement. This may include monitoring water usage, chemical inputs, or equipment efficiency.

Automated Systems: Integrate smart technologies (such as smart irrigation systems, automated maintenance equipment, and digital communication tools) to reduce human error and optimize routine tasks.

Process Improvement Strategies

Standardizing Procedures: Develop and regularly update SOPs for all critical processes. This ensures consistency and helps staff understand the best practices for each task.

Continuous Improvement: Encourage a culture of lean management where staff are empowered to suggest and test new ideas that shave time and cost from operations.

Resource Reallocation: Regularly review staffing, equipment, and budgets to ensure that resources are aligned with the areas of highest impact. Efficiency gains here translate directly into improved operational performance.

C. Aligning Operational Efforts with Strategic Vision

Effective leaders ensure that operational efficiency is not an isolated effort but is aligned with the overall strategic vision of the organization. This means:

Setting Clear Performance Metrics: Define key performance indicators (KPIs) for various operations. Regularly track and review these metrics to understand progress and pinpoint areas for improvement.

Engaging Staff in Efficiency Initiatives: Involve team members in brainstorming sessions to tap into their on the ground insights. When staff see that their suggestions lead to tangible efficiency improvements, it increases buy in and operative morale.

Leveraging Technology for Improved Coordination: Integrate communication platforms and management software that foster coordination between departments, ensuring that all operational elements are working in concert.

D. Real World Example

A distinguished golf course management team implemented a digital resource management system that centralized data on everything from water usage to staffing schedules. Through precise monitoring and routine process audits, they identified redundancies and bottlenecks that had previously gone unnoticed. Streamlined procedures not only led to a 15% reduction in operational costs but also improved response times to maintenance issues, which in turn increased overall course satisfaction among both golfers and staff.

E. Diagram Idea

A process map detailing the "Operational Efficiency Cycle" can be very helpful:

Identify & Analyze Processes: (Collect data on current processes)

Implement Improvements: (Introduce automated systems and standardized procedures)

Monitor Outcomes: (Use KPIs and feedback loops)

Refine and Repeat: (Continuous improvement cycle) This diagram would visually reinforce how incremental enhancements contribute to overall efficiency.

F. Checklist Course Operational Efficiency

Develop and continually refine standard operating procedures (SOPs) across all operational areas.

Implement digital tools and management software to centralize data and streamline processes.

Set, monitor, and review KPIs related to key operational functions.

Engage staff in efficiency improvement initiatives and document their feedback.

Use resource reallocation strategies to ensure optimal use of manpower, equipment, and budget.

Hold regular process audits to identify and eliminate bottlenecks.

Foster a culture of continuous improvement by celebrating efficiency wins and learning from inefficiencies.

Conclusion: Leadership as the Catalyst for Success

Becoming an effective leader in golf course management transcends possessing technical knowledge it demands the ability to inspire, resolve conflicts, and continually drive operational excellence. By leading staff effectively, resolving conflicts constructively, and ensuring that all operational aspects run efficiently, you create an environment where both the team and the course flourish. Leadership is not static; it is an ongoing journey that involves continuous learning, adaptation, and self reflection.

As you build your leadership toolkit, remember that every conversation, every decision, and every efficiency improvement contributes to a culture of excellence. This integrated approach will not only raise the operational standards of your course but also foster an environment where both staff and golfers feel engaged, respected, and motivated. In this way, effective leadership becomes the invisible force that propels the entire organization toward sustained success.

Final Summary Checklist for Becoming an Effective Leader

Leading Staff Effectively:

Communicate a clear vision and set transparent expectations.

Delegate tasks strategically and empower your team.

Engage in regular training, feedback sessions, and recognition programs.

Build a culture of open communication and trust.

Conflict Resolution and Golfer Satisfaction:

Establish formal conflict resolution protocols.

Train staff in conflict management and customer service.

Implement feedback mechanisms to address and preempt issues.

Use early intervention strategies to maintain a positive environment.

Course Operational Efficiency:

Develop and adhere to standard operating procedures.

Embrace technology and digital tools for process optimization.

Set clear KPIs and engage staff in continuous improvement.

Conduct regular process audits and reallocate resources as needed.

Final Thoughts

The journey to becoming an effective leader in golf course management requires as much dedication to people as it does to process. By nurturing a motivated, well drilled team, resolving conflicts with empathy and foresight, and relentlessly pursuing operational excellence, you set the stage for a thriving, resilient organization. Your leadership style directly impacts daily operations, influences the work culture, and shapes the satisfaction of both staff and golfers. Embrace the dual responsibility of managing people and processes, and remember that the best leaders are always learning, adapting, and inspiring others to achieve greatness.

With the strategies, real world examples, and actionable checklists provided in this chapter, you are well equipped to elevate your leadership skills and drive your golf course toward both operational success and a legacy of excellence.

Chapter 14: Business & Budgeting Strategies

Introduction

In golf course management, sound business and budgeting strategies are as crucial as the physical maintenance of the course. Financial planning for course improvement, building strong sponsorship and networking channels, and effective cost management in turf maintenance work together as the backbone of a successful, sustainable operation. With the rising costs of equipment, labor, and maintenance inputs, it is imperative to develop robust financial plans that not only allocate resources efficiently but also generate additional revenue streams. This chapter explores these three key areas:

Financial Planning for Course Improvement: Setting the stage for long term growth, these strategies help you develop realistic budgets, secure capital funding, and plan for major capital projects.

Sponsorships and Networking: Effective networking and sponsorship programs not only provide additional funds but also enhance your course's brand value and community standing.

Cost Management in Turf Maintenance: Smart cost management measures help reduce waste, optimize resource use, and ensure that every dollar spent contributes to high quality turf and improved playability.

By integrating these strategies into your business approach, you can maximize the return on investment of every project while safeguarding your course's financial health.

Financial Planning for Course Improvement

A. Establishing a Financial Vision

Long range capital planning and budgeting are critical for ensuring that your course remains competitive and well maintained. Financial planning for course improvements involves:

Identifying Needs: Conduct a thorough review of your course to determine which areas need improvement be it upgrading irrigation systems, renovating clubhouse facilities, or modernizing turf management equipment.

Setting Priorities: Not all projects can be undertaken at once. By prioritizing improvements based on immediate operational needs and long term strategic goals, you create a roadmap that connects short term fixes with long range investments.

Developing a Master Budget: Create a comprehensive budget outlining capital expenditures (CapEx), operational expenditures (OpEx), and projected revenue streams over a given timeframe. This budget should include contingencies for unexpected expenses while reflecting realistic growth projections.

Real World Example: For instance, one high end course faced declining membership due to outdated facilities. The superintendent worked closely with the board to develop a 5 year master plan that allocated funds for gradual renovations starting with the greens and later expanding to clubhouse upgrades. Through careful budget planning and phased implementation, the improvements not only enhanced the course's aesthetics but also attracted new members, resulting in higher overall revenue.

B. Capital Improvement Funding Options

Securing the necessary funding to finance course improvements can come from several sources:

Bank Loans and SBA Loans: Traditional bank financing can be a viable route if you have a solid credit history and a well crafted business plan. SBA loans, in particular, may offer favorable terms for smaller operations.

Reserve Funds: Many courses maintain reserve funds pooled over time to cover capital improvements. A disciplined approach to budgeting throughout the year can help ensure there is sufficient cash available when needed.

Asset Liquidation: In certain cases, underutilized assets can be sold to generate immediate cash. This might include older machinery or parts of the course that no longer provide significant value.

Public Private Partnerships (PPPs): For larger projects, forging partnerships with local businesses or government entities can share the financial risk while bolstering community support.

Diagram Idea: Imagine a flowchart that starts with "Identify Improvement Areas" and branches to

"Develop Master Budget," then to multiple funding sources (Bank Loans, Reserve Funds, Asset Liquidation, PPPs), with arrows showing due diligence steps and approval processes. This visualization helps simplify the intricate financing journey.

C. Tools and Technologies to Support Financial Planning

Financial Management Software: Implement robust financial systems that track operating costs, expenditures, revenues, and forecasting data. Software that integrates budgeting with historical maintenance data can help pinpoint areas where costs can be reduced or where investment will yield the highest return.

Dashboard Reporting: Develop or adopt dashboards that provide real time financial metrics. Visualizations such as graphs and heat maps can help you quickly identify trends and pivot strategies when necessary.

D. Checklist Financial Planning for Course Improvement

Conduct a thorough needs assessment across all areas of the course.

Develop a prioritized list of improvement projects.

Create a multiyear master budget covering both CapEx and OpEx.

Identify potential funding sources and prepare necessary documentation for bank or SBA loans.

Establish reserve funds and monitor cash flow to ensure sufficient liquidity.

Invest in financial software and dashboard systems to track ongoing performance.

Regularly review and update the budget based on project milestones and market conditions.

Sponsorships and Networking

A. The Power of Strategic Partnerships

Sponsorships and networking play a vital role in supporting course improvements and day to day operational expenses. By forging strategic partnerships, golf courses can access external funds, enhance their brand image, and create additional revenue streams, all while building lasting relationships within the industry. Sponsorships may come from local businesses, national brands, or charitable organizations, often in exchange for advertising, hosting events, or co branded marketing campaigns.

B. Crafting a Compelling Sponsorship Proposition

To attract sponsors successfully, it is essential to:

Understand Your Audience: Identify the demographic and market segments that frequent your course. Tailor your sponsorship proposal to demonstrate how aligning with your course can benefit the sponsor by reaching their ideal clientele.

Showcase Your Unique Value: Emphasize the unique attributes of your course its history, environmental commitment, strategic location, and quality of play. Data such as membership numbers, event attendance, and media reach can enhance your proposal.

Offer Tiered Sponsorships: Develop multiple sponsorship levels that match different budgets and expectations. For example, exclusive title sponsorships may come with significant branding opportunities, while smaller packages might include banners, digital mentions, or event booths.

Leverage Events and Tournaments: Golf tournaments and charity events are excellent platforms to showcase sponsors. Use these events to build ongoing relationships and drive high levels of engagement with the sponsor's target audience.

Real World Example: A mid size club in a suburban community successfully boosted its capital improvement budget by actively pursuing sponsorships for its annual tournament. By offering tiered packages that included branding opportunities on digital platforms, the club raised a significant sum from local and regional businesses. This additional funding not only helped finance minor course upgrades but also enhanced the club's reputation within the local business community, fostering longer term partnerships.

C. Building an Effective Networking Strategy

Attend Industry Conferences: Conferences and trade shows are excellent venues to meet potential sponsors, network with peers, and discover emerging trends within the industry.

Engage Local Business Leaders: Host networking events on site, invite community business leaders for tours of the course, and discuss mutually beneficial opportunities.

Utilize Social Media and Digital Marketing: Digital platforms provide a broad reach. Regularly update your course's social media channels, share success stories, and create compelling content that highlights your course's unique strengths.

Maintain Regular Communication: Sponsorship is an ongoing relationship. Regular updates, newsletters, meetings, and invitations to events keep sponsors engaged and increase the likelihood of renewal or expansion of sponsorship deals.

D. Diagram and Visualization

Diagram Idea: Illustrate a "Sponsorship Ecosystem" diagram that connects your golf course to various sponsorship opportunities. For example, create circles for "Event Sponsorship," "Digital Exposure," "Physical Branding," and "Community Partnerships," with arrows indicating the mutual benefits and value exchanges between the course and sponsors.

E. Checklist Sponsorships and Networking

Develop a detailed sponsorship proposal that includes audience demographics, branding opportunities, and benefit tiers.

Identify potential sponsor leads and categorize them based on size, industry, and alignment with your course reputation.

Plan and host networking events or sponsor appreciation days to build and reinforce partnerships.

Utilize digital marketing and social media to enhance your sponsorship appeal.

Establish regular communication channels (newsletters, reports) to update sponsors on the course's progress and upcoming opportunities.

Measure and document sponsorship outcomes, adjusting the proposal based on feedback and ROI analysis.

Cost Management in Turf Maintenance

A. The Importance of Controlling Turf Maintenance Costs

Efficient turf maintenance is the lifeblood of golf course operations, but it is also one of the largest ongoing expenses. Effective cost management involves balancing quality maintenance and sustainability with operational budgeting and strategic cost reduction. By streamlining processes, adopting advanced technologies, and controlling input costs, you can achieve significant savings while maintaining excellent turf quality.

B. Strategies for Cost Management

Optimize Labor Management:

Training and Team Efficiency: Invest in staff training to ensure that all team members are working efficiently. Cross training can maximize productivity and reduce labor redundancies.

Scheduling and Workload Balancing: Utilize scheduling software to ensure that manpower is allocated optimally throughout the season. Planning work during off peak hours can reduce overtime costs and enhance team efficiency.

Equipment and Resource Management:

Preventative Maintenance: Prioritize regular equipment servicing and repair schedules (as outlined in earlier chapters) to prevent costly downtime and extend the lifespan of your machinery.

Smart Purchasing: Consider bulk purchasing and long term supplier contracts for chemicals, fertilizers, and equipment parts. These arrangements often yield better pricing and more consistent product availability.

Adopt Energy Efficient Practices: Lowering energy consumption in irrigation, lighting, and other operational areas directly translates into cost savings over time.

Advanced Technology Adoption:

Digital Tools and Software: Implement financial management and budgeting tools that help track expenses, monitor cost trends, and forecast future spending needs. Data analytics can flag areas of wastage and inefficiency.

Precision Maintenance Technologies: Leverage technology like GPS mapping, drone surveys, and soil sensors (as discussed in previous chapters) to target maintenance efforts precisely. This reduces chemical and water waste and improves overall maintenance efficiency, translating to lower costs over time.

C. Real World Example

A golf course located in a competitive suburban market recognized that unanticipated equipment downtime and inefficient labor scheduling were the primary contributors to escalating maintenance costs. By adopting a digital maintenance management system and restructuring labor schedules to better align with course needs, the course reduced its maintenance expenses by over 20% within a single season. Regular reviews of chemical usage and targeted applications based on sensor data further contributed to overall cost reductions.

D. Diagram and Visualization

Diagram Idea: Develop a cost flow diagram that breaks down the major expense categories in turf maintenance (labor, chemicals, equipment, and energy). Overlay this with potential savings opportunities based on implementing technology, preventative maintenance, and optimized scheduling. A “before and after” comparison chart can be particularly impactful in visualizing cost savings over time.

E. Checklist Cost Management in Turf Maintenance

Regularly review and update maintenance budgets using digital financial tools.

Implement preventative maintenance schedules to reduce equipment downtime.

Optimize labor schedules and cross train staff to improve efficiency.

Monitor chemical and fertilizer applications to ensure precision targeting.

Invest in energy efficient and precision maintenance technologies.

Establish supplier contracts and invest in bulk purchasing when feasible.

Conduct periodic audits of maintenance costs and adjust strategies accordingly.

Use data analytics to identify areas of wastage and opportunities for process improvement.

Conclusion: Building a Sustainable Business Model

For golf course superintendents, effective business and budgeting strategies are not just about controlling costs they're about building a resilient model that supports continuous improvement and long term growth. By strategically planning financial resources for course improvements, forging strong relationships through sponsorships and networking, and managing turf maintenance costs with precision, you can ensure your operations remain competitive and sustainable.

These strategies are interdependent: sound financial planning creates the foundation for major course improvements; a robust sponsorship network provides additional funding and enhances brand credibility; and disciplined cost management optimizes daily operations, freeing up resources for strategic investments. Together, these elements create a cycle of continuous improvement, driving your course toward increased profitability, sustainability, and long term operational success.

Remember, the key to effective budgeting isn't just saving money it's investing wisely in the future of your course. With a comprehensive approach supported by detailed checklists, real world examples, and innovative diagram tools, you are well equipped to navigate the financial challenges of golf course management and make strategic decisions that benefit both the short term operations and the long term vision of your facility.

Final Summary Checklist for Business & Budgeting Strategies

Financial Planning for Course Improvement:

Conduct a comprehensive needs assessment of course infrastructure and maintenance.

Develop a multiyear master budget that integrates CapEx, OpEx, and contingency plans.

Identify and secure funding sources (bank loans, reserve funds, PPPs, asset liquidation).

Invest in financial management software and dashboard reporting for real time tracking.

Sponsorships and Networking:

Develop a tailored sponsorship proposal highlighting your course's unique selling points.

Create tiered sponsorship packages and build relationships with local and national sponsors.

Organize networking events and maintain regular communication with current and potential sponsors.

Leverage digital marketing to amplify sponsorship benefits and reach target audiences.

Cost Management in Turf Maintenance:

Optimize labor management through training and scheduling software.

Implement preventative maintenance protocols to reduce equipment downtime.

Use technological tools (drones, sensors, digital analytics) for targeted inputs.

Benchmark costs regularly and adjust purchasing, energy use, and chemical applications as needed.

Final Thoughts

A well crafted business and budgeting strategy is the lifeblood of a successful golf course operation. The decisions you make today about how to allocate resources, engage sponsors, and control costs will have long lasting effects on your course's quality, reputation, and profitability. By embracing a proactive approach to financial planning, cultivating meaningful sponsorships through effective networking, and remaining vigilant in managing turf maintenance costs, you not only safeguard the short term viability of your operations but also lay the groundwork for sustained long term growth.

As you move forward, remember that financial stewardship in golf course management is an evolving discipline one that requires continuous learning, adaptation, and a forward thinking mindset. By leveraging the strategies, tools, and techniques outlined in this chapter, you can transform your financial challenges into opportunities for innovation and lasting success.

Chapter 15: Career Development & Industry Networking

Introduction

In the competitive world of golf course management, technical expertise is only one piece of the puzzle. The path to professional success and to truly standing out in your field requires a commitment to continuous learning, thoughtful career development, and strategic networking. As a golf course superintendent or manager, your growth as a leader and your ability to forge strong industry connections can directly influence your career trajectory and the operational success of your course. This chapter dives into three critical areas for career development: obtaining professional certifications and engaging in industry learning; participating in turf conferences and trade shows; and cultivating lasting relationships with suppliers and partners.

Whether you are just starting in the industry or are a seasoned professional, investing in your personal and professional development is essential. These efforts can lead to enhanced credibility, increased job satisfaction, and opportunities for advancement, all of which have downstream benefits for the courses you manage. In the sections that follow, you will find detailed strategies, real world examples, and actionable checklists that serve as a roadmap for climbing the career ladder and building a strong industry network.

Professional Certifications and Industry Learning

A. The Importance of Continuing Education

In an industry marked by rapid technological advancement and evolving best practices, staying current through continuous learning is not just beneficial it's essential. Professional certifications and targeted training programs offer a structured pathway to acquiring new skills, validating existing expertise, and distinguishing yourself from the competition. These credentials provide tangible proof of your commitment to excellence and can open doors to advanced roles, higher salaries, and leadership opportunities.

Key Areas of Focus:

Technical Expertise: Obtaining certifications in agronomy, turf management, or environmental science ensures you can leverage the latest techniques and tools.

Leadership and Management: Courses and certifications in business administration, project management, or finance (often available through specialized golf management programs) bolster your ability to lead teams and manage large scale projects effectively.

Emerging Technologies: With the increasing importance of automation and precision management, certifications focused on smart irrigation systems, GPS mapping, or data analytics provide a competitive edge in today's market.

B. Certification Programs and Industry Organizations

There are several reputable organizations and academic institutions that offer certifications tailored to golf course management:

PGA of Canada and PGA of America: These organizations offer certifications that encompass both the technical and business aspects of golf management, ensuring you meet rigorous industry standards.

Golf Management Programs: Numerous institutions, such as Humber College and McMaster Continuing Education, provide specialized diplomas and certificates in golf course management that include hands on training and field experiences.

Industry Specific Certifications: Programs from associations like the United States Golf Teachers Federation (USGTF) or the Golf Management Institute of Canada (GMIC) serve to validate your expertise and enhance your professional profile.

C. Engaging in Continuous Learning

Beyond formal certifications, the culture of continuous learning is sustained by:

Online Courses and Webinars: Numerous online platforms offer modules on the latest trends in turf management, sustainable practices, and emerging technologies.

Industry Publications and Journals: Regularly reading industry magazines, research papers, and case studies can keep you informed about the latest innovations and industry challenges.

Mentorship Programs: Developing relationships with more experienced industry leaders can provide invaluable real world insights and career guidance.

Real World Example: A superintendent in a highly competitive metropolitan region pursued an advanced certification in sustainable turf management through the PGA of Canada. By coupling this credential with a series of online courses in leadership and data analytics, the superintendent not only improved his technical skill set but also garnered the respect of peers and upper management leading to a promotion and broader strategic responsibilities.

D. Diagram and Visualization

Diagram Idea: Develop an infographic entitled "Your Path to Professional Excellence," starting with foundational certifications in turf management, branching into leadership and specialized technology courses, and culminating with ongoing professional development (webinars, industry publications, mentorship). This visual can serve as a roadmap, illustrating clear career milestones and the interconnected benefits of continuous learning.

E. Checklist Professional Certifications and Industry Learning

Identify the core areas where additional certification would enhance your expertise (e.g., turf science, sustainability, leadership).

Research and select accredited programs or institutions offering the needed certifications.

Develop a learning calendar that includes formal coursework, online courses, webinars, and industry conferences.

Subscribe to leading industry publications and join professional associations.

Seek mentorship opportunities with seasoned professionals in your field.

Regularly review new courses and emerging specializations to stay ahead of industry trends.

Document your certifications and professional development achievements for career advancement reviews.

Attending Turf Conferences and Trade Shows

A. The Value of Industry Events

Turf conferences, trade shows, and industry events provide unparalleled opportunities to learn about new technologies, discover innovative products, and network with industry experts. They serve as a hub where knowledge meets opportunity a place where the latest trends are showcased and where real world applications are discussed. Attending these events can be instrumental in:

Learning: Gaining insights into the latest research, technologies, and best practices in turf management.

Networking: Connecting with vendors, peers, thought leaders, and potential collaborators.

Showcasing: For those with groundbreaking ideas or products, these events offer an ideal platform to demonstrate innovative solutions and build brand recognition.

B. Maximizing Conference and Trade Show Benefits

When planning to attend industry events, consider the following strategies:

Pre Event Preparation:

Set Clear Objectives: Define your goals for attending whether it's learning about emerging technologies, meeting suppliers, or exploring partnership opportunities.

Research the Agenda: Look up keynote speakers, panel discussions, and exhibitor lists ahead of time. Identify sessions that are most relevant to your current challenges and future goals.

Schedule Meetings: Reach out to vendors or key industry professionals in advance to set up one on one meetings during the event.

Active Participation:

Engage and Ask Questions: Don't hesitate to participate during sessions. Ask questions and share your own experiences; this not only increases your visibility but also enriches your learning.

Attend Networking Sessions: Use coffee breaks, luncheons, and organized networking events to interact with other industry professionals.

Collect and Organize Materials: Bring business cards, brochures, and notepad apps on your phone so you can capture important details effortlessly.

C. Post Event Follow Up

Review Your Notes: Soon after the event, review and organize your notes to identify key takeaways and potential opportunities.

Follow Up Meetings: Reach out to contacts you met to further discuss potential collaborations or to simply thank them for the insights shared.

Integrate Learning: Consider how you can implement the innovative practices and strategies discussed at the event into your own operations. Share these insights with your team to boost overall performance.

Real World Example: A superintendent attending the Canadian Golf Course Management Conference discovered a new organic pest management technology during a breakout session. By networking with the company representatives and attending subsequent product demonstrations, the superintendent not only implemented the technology on his course which resulted in a 15% reduction in pesticide use but also built a relationship that led to favorable pricing and technical support for future upgrades.

D. Diagram and Visualization

Diagram Idea: Create a "Conference Networking Cycle" diagram that includes key phases:

1. Pre Event Planning (goal setting, scheduling meetings),

2. Active Participation (attending sessions, engaging in networking),

3. Post Event Follow Up (reviewing notes, scheduling follow up discussions),

4. Integration of Insights (applying new ideas to your course). This can be a visually appealing way to illustrate how conferences and trade shows can turn into real operational improvements.

E. Checklist Attending Turf Conferences and Trade Shows

Set clear objectives for each event (e.g., learning, networking, sourcing new technologies).

Research and select relevant conferences and trade shows well in advance.

Prepare business cards, digital contact methods, and a professional pitch.

Schedule meetings with key contacts ahead of the event.

Participate actively during sessions and use every opportunity to ask questions.

Collect all relevant materials and make detailed notes.

Follow up promptly with new contacts post event.

Document actionable insights and evaluate how they can be integrated into your operations.

Relationship Building with Suppliers and Partners

A. The Strategic Importance of Strong Relationships

In the world of golf course management, the relationships you build with suppliers, vendors, and strategic partners can have a profound impact on your operation's success. These relationships go beyond transactional exchanges; they become the cornerstone of a reliable supply chain, innovative collaborations, and cost effective solutions. Successful relationship building can lead to negotiated discounts, priority support during emergencies, and insider access to new technologies and products.

B. Strategies for Cultivating Supplier and Partner Relationships

Establish Trust and Communication:

Regular Meetings: Schedule regular check ins with suppliers and partners. Whether through phone calls, face to face meetings, or virtual conferences, consistent communication reinforces trust.

Transparency: Be open about your needs, expectations, and challenges. Clear communication helps suppliers better tailor their offerings to your requirements.

Timely Payments and Fair Deals: Ensure that invoices are paid promptly and that the terms of your agreements are transparent and mutually beneficial. Consistently honoring your commitments promotes long term loyalty.

Collaborative Problem Solving:

Joint Planning Sessions: Work with your suppliers to anticipate seasonal challenges and plan for stock or service needs in advance.

Feedback Mechanisms: Provide constructive feedback that helps suppliers improve their products or services. Regular reviews and feedback sessions create a partnership dynamic where both parties feel invested in mutual success.

Personal Touch: Small gestures such as thank you notes or invitations to industry events can strengthen relationships and create lasting bonds.

C. Leveraging Digital Tools for Supplier Management

In today's connected world, digital platforms can streamline relationship management:

CRM Systems: Use customer relationship management systems to track communication history, upcoming contract renewals, and special deals. This organized approach ensures that no detail is overlooked.

Vendor Portals: Encourage suppliers to use online portals where order statuses, feedback, and performance metrics are transparent. This promotes accountability and efficiency.

Social Media Engagement: Follow key suppliers on professional networks such as LinkedIn, and partake in industry forums to stay abreast of new developments and opportunities.

Real World Example: A superintendent at a prestigious golf course regularly organized quarterly "supplier roundtables," where major vendors would come together for a casual discussion about upcoming needs and challenges. This initiative not only fostered better communication but also led to bulk order discounts and a collaborative approach to addressing critical course challenges during peak seasons.

D. Diagram and Visualization

Diagram Idea: Design an infographic titled "Building Strategic Supplier Relationships." The diagram might include:

Key Touchpoints: Initial contact, regular check ins, collaborative problem solving, performance reviews, and joint planning sessions.

Benefits: Improved pricing, reliable service, advanced product access, and emergency support.

Feedback Loop: How ongoing communication and performance measurement feed back into strengthening the relationship.

E. Checklist Relationship Building with Suppliers and Partners

Create a list of key suppliers and strategic partners and review the terms of your current agreements.

Schedule regular meetings monthly or quarterly to discuss performance and future needs.

Implement a CRM system to track interactions and follow up on commitments.

Provide timely, constructive feedback and recognize supplier excellence.

Explore joint ventures or collaborative projects to improve operational efficiency.

Utilize digital tools to enhance transparency and accountability in all transactions.

Maintain an updated contact database and ensure open lines of communication.

Conclusion: Paving the Way for a Proactive, Networked Career

Building a thriving career in golf course management and advancing the operational success of your course requires ongoing professional development, proactive attendance at industry events, and a systematic approach to relationship building with suppliers and partners. This chapter has outlined strategies and best practices that not only foster personal and professional growth but also ensure that your operation remains on the cutting edge. By investing in industry certifications and continuous learning, actively engaging in turf conferences, and nurturing strong supplier relationships, you lay the foundation for a long and prosperous career and create a robust network that supports your organization in every season.

The relationships you build and the knowledge you acquire can create a ripple effect, driving innovation, operational efficiency, and increased profitability. In a world where both technological advancements and relationship dynamics are rapidly evolving, a proactive, well networked career strategy is your key to unlocking new opportunities and staying ahead of the curve.

Final Summary Checklist for Career Development & Industry Networking

Professional Certifications and Industry Learning:

Identify and pursue relevant certifications in turf management, sustainability, and leadership.

Enroll in online courses, webinars, and advanced training programs.

Stay current with industry publications and research.

Seek mentorship and join industry associations to broaden your perspective.

Attending Turf Conferences and Trade Shows:

Set defined objectives for attending major industry events.

Research and select conferences and trade shows that align with your professional goals.

Schedule meetings with potential industry contacts ahead of time.

Engage actively during sessions and capture actionable insights.

Follow up with new contacts and integrate learning into your operations.

Relationship Building with Suppliers and Partners:

Maintain regular communication with key industry suppliers.

Establish a systematic approach to giving and receiving feedback.

Use digital tools to track interactions and performance.

Organize joint planning sessions and networking events.

Develop long term contracts that benefit both parties.

Final Thoughts

Your career development and industry networking efforts are integral to your personal success and the success of your golf course. By embracing a mindset of continuous learning, proactively participating in industry events, and building strong, collaborative relationships with suppliers, you create opportunities to innovate and lead. Remember, every certification earned, every conference attended, and every relationship nurtured is a stepping stone toward realizing your full potential as a leader in golf course management.

Invest in your future, keep your network robust, and leverage every opportunity to learn and grow. The connections you build and the knowledge you accumulate will not only enhance your operational effectiveness but also open doors to new avenues in an ever evolving industry. Strive to be a proactive force in your career and a trusted partner in your industry your journey toward excellence is continuous, and every step you take today lays the groundwork for tomorrow's success.

Conclusion

As you reach the final pages of this book, it is worth pausing to reflect on the multifaceted journey we have embarked upon together. In today's ever evolving landscape of golf course management, success is defined by the integration of technical expertise, forward-thinking operational strategies, and a commitment to sustainable practices. This book has provided a comprehensive roadmap from the intricate details of turf science to the innovative strides in automation and precision management; from robust business and budget strategies to the indispensable art of leadership and professional networking.

A Journey Through Innovation and Stewardship

We began our journey by laying a strong foundation in the principles of sustainable turf management and course design. You learned how the bedrock of a course its soil, its water system, its mowing regimes directly influences everything from turf performance to playability. The early chapters emphasized the importance of data-driven decisions: regular soil testing, strategic fertilization, carefully calibrated irrigation routines, and precise mowing techniques. These practices are not mere routine tasks, but rather the building blocks of a resilient course that stands up to both natural challenges and the rigors of play.

As we moved forward, the focus shifted to equipment and infrastructure maintenance. Robust maintenance strategies, including routine equipment servicing, thoughtful infrastructure upgrades (like bunker drainage repairs and winterization practices), and preventative maintenance scheduling, are critical for reducing downtime and ensuring long term operational excellence. These chapters helped illustrate that proactive planning and attention to detail can result in significant cost savings and enhanced turf performance, even when the elements are less than forgiving.

The discussion evolved further as we tackled the challenges of managing seasonal transitions. From the vigorous recovery following winter dormancy to the precise interventions required in early spring, through the intense heat and drought of summer, and into the structured planning necessary for the fall transition and winter dormancy, every chapter provided actionable strategies to keep your course in peak condition year round. Each seasonal segment underscored a simple yet profound principle: that effective course management is a continuous, cyclical process of planning, monitoring, and adapting.

Equally significant were our explorations into the future of turf management. The integration of advanced technologies such as autonomous mowers, robotic systems, drone based turf monitoring, artificial intelligence driven disease detection, and soil sensor networks signals a new era in golf course management. However, these innovations also raise important practical considerations. We acknowledged that while such technologies offer tremendous potential for precision and efficiency, their advanced costs generally mean that only high end courses with deep budgets might adopt them fully in the near term. For many, these innovations serve as aspirational benchmarks outcomes that inspire gradual, incremental steps toward embracing technology in ways that balance cost with benefit.

On the leadership and business fronts, we have seen that financial acumen, effective team management, and robust networking are as decisive for success as technical proficiency. The chapters on leadership, budgeting, and career development illustrated that strong, visionary leadership creates an environment where every team member is motivated, every conflict is resolved with empathy and clarity, and every operational process is optimized for efficiency and sustainability.

Transforming Challenges into Opportunities

Throughout the book, the recurring theme is one of transformation from the grass on your course to the very management practices that sustain it. The challenges you face, whether they are climatic extremes, equipment downtime, or tight budgets, can be turned into opportunities for meaningful improvement, innovation, and growth. The integration of sustainable practices ensures that every resource is used responsibly, paving the way for long term success both financially and environmentally.

Embracing change is not always easy. It demands that you remain curious, steadfast, and resilient in the face of unpredictable challenges. Yet, every strategy, every checklist, and every technological innovation discussed in these pages has been chosen not only for its potential to improve turf quality but also to empower you as a leader in your field. The future of golf course management requires a harmonized approach where operational efficiency, environmental stewardship, and technological advancement work hand in hand.

A Call to Action

As you close this book, I encourage you to take the insights and strategies you have acquired and apply them with deliberate intent. Evaluate your current practices, identify areas where incremental changes can yield significant improvements, and set measurable goals for the future. Whether you are refining your irrigation practices, enhancing your maintenance scheduling, or taking that first step toward adopting digital tools and automation, remember that every small improvement contributes to a larger cycle of excellence.

Continue learning whether through industry certifications, conferences, or everyday experiences and build networks that help you remain at the forefront of innovation. Your commitment to professional growth and environmental stewardship is not only an investment in your career but also in the sustainable future of the golfing community.

Final Thoughts

In the end, this book is more than a technical guide or a business manual it is a manifesto for continuous improvement in an industry that is as much about nature as it is about nurture. It challenges you to balance precision with adaptability, to combine data driven decisions with human insight, and to view every challenge as a stepping stone toward a more sustainable and successful future.

May the strategies, tools, and insights shared here serve as a constant resource and a source of inspiration as you lead your team, manage your course, and navigate the exciting future of golf course management. Embrace the journey, be the catalyst for change, and remember: every blade of grass, every drop of water, and every decision counts.

Thank you for joining me on this journey toward excellence in golf course management. Cheers to a future of greener fairways, innovative solutions, and sustainable success.

www.ingramcontent.com/pod-product-compliance
Lightning Source LLC
LaVergne TN
LVHW050929210926
848482LV00013B/575
9781069632449